T0289176

Magnetometry
for Archaeologists

Geophysical Methods for Archaeology

Series Editors:
Lawrence B. Conyers, University of Denver,
Kenneth L. Kvamme, University of Arkansas

The Geophysical Methods for Archaeology series aims to introduce field archaeologists and their students to the theory and methods associated with near-surface geophysical data collection techniques. Each book in this series will describe one of these commonly used noninvasive techniques, its applications, and its importance to archaeological practice for the nonspecialist.

Volume 1: *Ground-Penetrating Radar for Archaeology*, Lawrence B. Conyers

Volume 2: *Magnetometry for Archaeologists*, Arnold Aspinall, Chris Gaffney, and Armin Schmidt

Magnetometry
for Archaeologists

Arnold Aspinall, Chris Gaffney, and Armin Schmidt

AltaMira
PRESS

A Division of Rowman & Littlefield Publishers, Inc.
Lanham • New York • Toronto • Plymouth, UK

ALTAMIRA PRESS
A division of Rowman & Littlefield Publishers, Inc.
A wholly owned subsidiary of The Rowman & Littlefield Publishing Group, Inc.
4501 Forbes Boulevard, Suite 200
Lanham, MD 20706
www.altamirapress.com

Estover Road
Plymouth PL6 7PY
United Kingdom

British Library Cataloguing in Publication Information Available

Library of Congress Cataloging-in-Publication Data

Aspinall, A.
  Magnetometry for archaeologists / Arnold Aspinall, Chris Gaffney, and Armin Schmidt.
    p. cm. — (Geophysical methods for archaeology ; v. 2)
  Includes bibliographical references and index.
  ISBN-13: 978-0-7591-1106-6 (cloth : alk. paper)
  ISBN-10: 0-7591-1106-5 (cloth : alk. paper)
  1. Magnetometry in archaeology. I. Gaffney, C. F. (Chris F.) II. Schmidt, Armin, 1961–
III. Title.

  CC79.M33A85 2008
  930.1—dc22   2007049378

Printed in the United States of America

™ The paper used in this publication meets the minimum requirements of American
National Standard for Information Sciences—Permanence of Paper for Printed Library
Materials, ANSI/NISO Z39.48-1992.

CONTENTS

LIST OF FIGURES, PLATES, AND TABLES

Figures

Plates

Tables

CONCEPTS

History of Magnetism

Magnets and magnetism are common concepts in today's world, and it is true that their effects are constantly with us and infinitely variable. They underpin many valuable tools for archaeological purposes, and have applications in provenance studies, dating, and prospecting for archaeological artifacts, features, and sites. The knowledge that natural materials have magnetic properties has a long history. Fragments of a black rock with the peculiar ability to attract other fragments of the same mineral were widely discovered and remarked upon in ancient times. When found in large quantities in Magnesia, Thessaly, then home of a Macedonian tribe called the Magnetes in what is now central Greece, the Greeks named the mineral *magnes lithos*, and later the Romans called it *magneta*, which eventually led to today's many related terms. Interestingly, no reference has been found in surviving Greek and Roman writings as to the directional properties of *magnes lithos* or *magneta* for its eventual use in a compass. Observations on these properties (in what the English, much later, called lodestone, or leading stone), appear first in Chinese literature before AD 400. By the eleventh century, Chinese seafarers were routinely using practical compasses that in turn came into widespread European use and manufacture, possibly via Arab traders, by the thirteen hundreds. In 1269, during the siege of an Italian city, Petrus Pereginus de Maricourt, a French military engineer, took time to write of pivoted and floating compasses in some detail. He described carrying out experiments on the interaction of iron fragments with lodestone specimens, noted their points of strongest magnetic activity, and named these their "poles," giving rise to the notion of polarity.

The next historic step in the scientific study of magnetic phenomena took place in 1600, when William Gilbert, the English physician and scientist, published his treatise in Latin, *De Magnete*. He gave full credit to the work of Petrus, then propounded a significant and, for its time, astounding attribute of the earth, that it is a giant magnet creating its own magnetic field. He referred to the field's angles of dip and declination as seen with a compass, and demonstrated that iron loses its faculty for magnetization when raised to red heat, then regains it when cooled. Rather than invigorating the investigation of magnetism, however, Gilbert's groundbreaking research appears to have dampened interest in magnetic studies for the next two hundred years, aside from a new practical use discovered in seventeenth century Scandinavia, where variations in the earth's field were measured by compass to locate iron deposits—the first recorded magnetic prospecting beneath the earth's surface.

In the nineteenth century, came a period of great innovation, and magnetism, by then regarded as a cornerstone of the classical theory of physics, was again researched with great vigor. In 1820, the Danish scientist Hans Christian Ørsted made the observation that an electric current could affect a magnetic needle, and was therefore a source of a magnetic field. By 1825, the existence of mutual magnetic forces between wires carrying electric currents was demonstrated by the French physicist André-Marie Ampère, who propounded the equivalence of a permanent magnet and a "magnetic shell" created by a defined loop of wire carrying an electric current. The concept of magnetism resulting from electric current circulating within a medium thus appeared for the first time, but without further explanation. Only with the development of atomic quantum theory at the beginning of the twentieth century was Ampère's proposal substantiated. Another practical advance of importance was made by Johann Carl Friedrich Gauss, the German mathematician and scientist who in 1832 hung a permanent bar magnet from a gold filament and succeeded in accurately measuring the strength of magnetic fields. Thus came the first practical magnetometer.

Today, measurement of the strength of the earth's magnetic field is commonplace, and many of the most basic facts and properties of magnetic materials are taken for granted. Nonetheless, in the remainder of this chapter we shall review our knowledge of magnetism and the highlights of its concepts. All are needed to understand the minuscule, but significant, anomalies in the earth's magnetic field that result from the effects of buried archaeological features and strata. Without this knowledge, it is impossible to fully understand the instruments that are used to detect these perturbations, or to interpret the meaning of their variations. Once this theoretical base has been considered, and the value of these techniques is understood, the study of magnetometer data becomes useful to identify the nature of individual features and to chart buried landscapes.

Magnetic Fields and Magnetic Induction

It is clear from the historical development of our knowledge of magnetism that there are two principal sources of magnetic effects: permanent magnets and electrical devices. The separate approaches to the subject via these two phenomena have, historically, led to disagreement and confusion among physicists. For our purposes, however, we will use only the outcome of discussions that are relevant to our goal of quantifying the results of geophysical magnetometry. To that end, the International System (SI) of units can be used.

A magnet attracts a piece of iron even when the two are not in contact, and, by use of a compass needle, a field-of-force pattern may be plotted in the immediate vicinity of the magnet, with a high concentration toward the ends of the magnet and an indication of a direction of the compass needle going from one end of the magnet to the other. The regions of concentration, and the observation of a compass needle aligning in the earth's magnetic field, lead to the concept of a north-seeking (N) pole at one end of the magnet and a south-seeking (S) pole at the other. It should be stressed that these are seeking poles pointing roughly toward the north and south geographic poles of the earth. Since experience has taught us that like poles repel one another and unlike poles attract, the implication is that the earth's magnetic south pole lies near its geographic north pole. As we shall see later, this has complications for conventions used in magnetic surveys. If we assign a pole strength $\pm p$ to each of these end locations, we can examine the properties of the magnet in terms of

forces. Effectively, if two individual poles could be separated by a distance
r, then there would be a force between them given by

$$F = C \cdot p_1 \cdot p_2/r^2.$$

Eq. 1.1.   Magnetic Force

There is an inverse-square law in this equation that is common to
other point fields of force, such as electric charges and gravitation. Here,
p_1 and p_2 are measures of the individual pole strengths, and the term C
is a constant dependent on the system of units used to define p_1 and p_2.
The direction of the force is along the line connecting the two poles.
That isolated poles cannot be achieved in practice (at this stage of our
scientific knowledge) does not negate the result, provided long magnetic
needles are used to minimize the effect of the opposite poles. This leads
to a definition of the force effect on a unit-N pole placed at a distance r
from a like pole of strength p. It is appropriate to refer to this as the
magnetic flux density B, which, being a force, is a vector having both
magnitude and direction, conventionally in the direction of motion of
the unit-N pole. Thus

$$B = C \cdot p/r^2.$$

Eq. 1.2.   Magnetic Flux Density

Following the standard view, these lines of magnetic flux will start on N-
poles and end on S-poles. The concept of lines of magnetic flux representing
the magnetic field was, in fact, conceived by James Clerk Maxwell
(1831–1879), the Scottish mathematician and theoretical physicist, who regarded them as endless lines indicating the direction and, by their concen-
traion, the magnitude of the flux density at any point. The lines passing
through the magnet from the S- to N-pole in fig. 1.1 emerge and return
along the directions observed with a compass needle, giving a realistic picture
of the magnetic flux lines pervading all matter. Magnetic flux is formally described in webers (Wb), and is simply analogous to a flow. Magnetic flux
density, however, is defined in terms of the magnetic flux passing normally
through a unit area. Its unit is weber per square meter (Wb m^2), more frequently expressed in tesla (1 T = 1 Wb m^2). In fact, the tesla is an impracti-
cally large unit for our purposes, and the subunits micro (μT = 10^{-6} T),
nano (nT = 10^{-9} T), or pico (pT = 10^{-12} T) are usually employed. With the

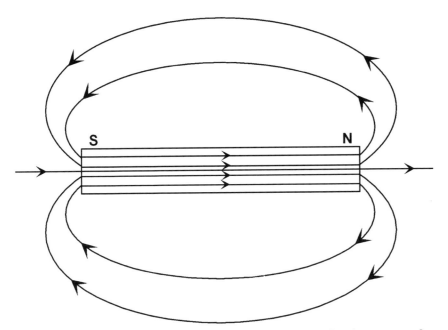

Figure 1.1. Magnetic flux lines in the vicinity of, and through, a bar magnet. Arrows indicate direction of field.

magnetic flux density of the earth varying from about 70 to 20 μT, magnetic archaeological features are typically measured in nanotesla.

The fact that any magnet comprises two poles of equal and opposite strength enables the flux density at any point, distant r_1 and r_2 from the two poles, to be calculated as the vector sum of

$$B_1 = +C \cdot p/r_1^2,$$

and

$$B_2 = -C \cdot p/r_2^2,$$

Eq. 1.3.

where the directions of B_1 and B_2 are given by the directions from the poles usually noted as r_1 and r_2, and the resultant B can be calculated by vector methods (fig. 1.2).

The behavior of a compass needle depends on the magnitude and direction of the magnetic field influencing the needle. A magnet of

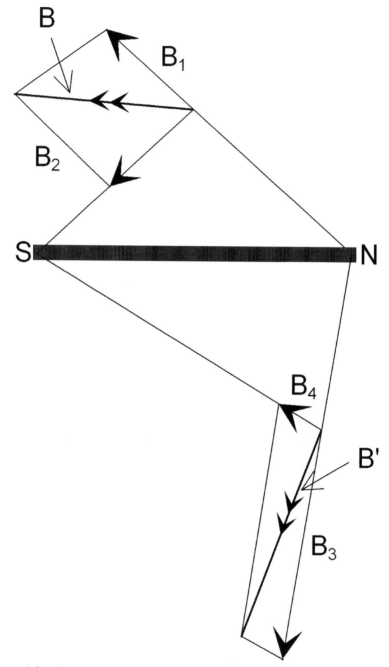

Figure 1.2.   Flux density lines of a bar magnet demonstrating the parallelogram
of vectors; B and B' are the resulting vectors.

length *l* and pole strength *p* will experience equal and opposite forces at its poles when placed in a field of uniform flux density. If the axis of the magnet is at an angle θ to the direction of the field, a turning couple, or torque, of magnitude

$$T = p \cdot l \cdot B \cdot \sin\theta$$
Eq. 1.4. Torque

will be experienced Additionally, *T* has vector properties in terms of the direction taken by the twist, and is referred to as clockwise or anticlockwise.

The product *p* · *l* is called the magnetic moment (*m*) of the magnet and defines the magnet's response to the applied torque. A freely suspended magnet can be shown to oscillate about the direction of a horizontal magnetic field with a frequency proportional to the square root of the flux density, thus providing a measure of its value. We will return to this property later.

The magnetic flux in the vicinity of a magnet is, clearly, the observed effect of a process occurring in the magnetic material, the cause of which resides in the structural properties of the material itself. Furthermore, the presence of any surrounding medium that contains the magnet must be accounted for. We may gain some insight into the cause of the magnetic flux by examination of the magnetic forces first noted by Ampère and Ørsted early in the nineteenth century. It can be shown experimentally that if a circular loop of wire of radius *r* in a vacuum (or air) carries an electric current *I* (ampere), a magnetic force field is observed such that at the center of the coil its direction, as seen with a compass needle, is perpendicular to the coil face. Its magnitude *H* is directly proportional to *I*, and inversely proportional to *r*:

$$H = I/2r.$$
Eq. 1.5. Magnitude

This force is a vector that is the cause of observed magnetic effects, and is given the unit termed ampere per meter (Am^{-1}). The observed magnetic flux lines that circulate as closed loops passing through the coil (fig. 1.3) bear a striking resemblance to the field created by a short bar magnet, assuming the magnet's flux lines pass through its length. By further comparison with

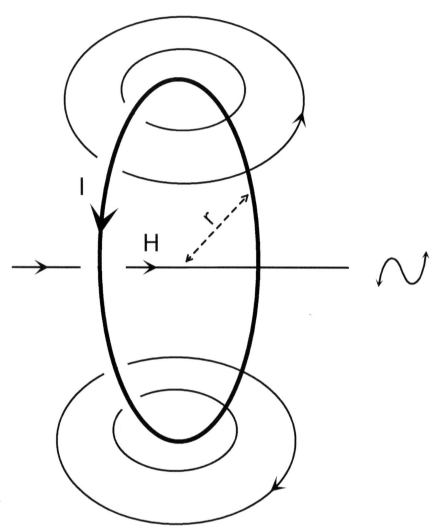

Figure 1.3. Magnetic field of force *H* created by a loop, radius *r*, carrying current *I*.

the magnet, we can ascribe a magnetic moment m to the loop, which is proportional to I and to the area A ($= \pi \cdot r^2$) of its face. Thus, with the unit ampere per square meter (Am^{-2}),

$$m = I \cdot A.$$

Eq. 1.6.   Magnetic Moment

We may regard *H*, the magnetizing field strength, as a cause of subsequent magnetic effects, and there must be a direct relationship between it and the resulting magnetic flux as measured by the magnetic flux density *B* in the containing medium. In a vacuum, we would expect *H* and *B* to be equal. However, because of the different origins of the units of the two, it is necessary to introduce a balancing constant of the form

$$B = \mu_0 \cdot H, \text{ (in a vacuum)}$$

Eq. 1.7. Magnetic Flux Density

where μ_0 is a constant known as the magnetic permeability of free space. Its value is $4\pi \cdot 10^{-7}\, T\,\text{m A}^{-1}$. If a medium is present, then its effect is to modify the magnetic flux passing through it such that

$$B = \mu_R \cdot \mu_0 \cdot H,$$

Eq. 1.8. Magnetic Flux Density

where μ_R is a number expressing the enhancement, or otherwise, from the presence of the medium. It is known as the relative magnetic permeability. This is an important property and may be a constant or a function of the magnetizing field.

Experience tells us that if susceptible material is placed in a magnetizing field, magnetism will be induced into it. This results in the creation of induced magnetic poles over faces perpendicular to the direction of *H*, the magnetizing field. The intensity of magnetization *M* created in the specimen by *H* may be expressed as the induced pole strength per unit area of face, but more frequently as the induced magnetic moment (*m*) per unit volume of specimen, i.e., as

$$M = m/l \cdot A,$$

Eq. 1.9. Intensity of Magnetization

where *l* is the specimen length and *A* the area of the pole face. For an arbitrarily shaped feature of volume *V*, the expression is simply $M = m/V$. As the unit of *m* is Am^2, this leads to the unit of *M* to be Am^{-1}, which is identical with the unit for H. Evidently the magnitude of M is dependent on H, and we may write

$$M = \kappa \cdot H,$$

Eq. 1.10. Intensity of Magnetization

where κ is a property we call magnetic susceptibility. As it is derived from the volume normalized magnetization M (i.e., total magnetic moment of the sample divided by its volume), it should—strictly—be called volume-specific magnetic susceptibility, indicating that it is a material constant and does not depend on the amount of sample investigated. Since M and H have the same units, κ is dimensionless. However, it is usual, though not mandatory, to annotate κ with SI units. This indicates that the value is measured with reference to the modern accepted standard. For laboratory measurements of susceptibility, it is common practice to take a particular mass of material when the magnetic value is expressed per unit mass. This quantity, mass-specific susceptibility χ, is then

$$\chi = \kappa/\rho,$$

Eq. 1.11. Mass Specific Susceptibility

where ρ is the bulk density of the medium. Its unit is then given as $m^3 \cdot kg^{-1}$. Since the total magnetic moment was normalized by the sample's volume to derive κ, and by the sample's mass to derive χ, the relationship between κ and χ is via the sample's bulk density.

Evidently there must be a relationship between the source H of magnetization, the induced magnetization M, and the overall resulting effect B, such that

$$B = \mu_0 \cdot (H + M),$$

Eq. 1.12. Magnetic Flux Density

or

$$= \mu_0 \cdot H \cdot (1 + \kappa).$$

Eq. 1.13. Magnetic Flux Density

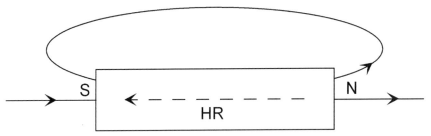

Figure 1.4.   Demagnetizing field *HR* within a rod magnetized along its axis.

We can thus identify $(1 + \kappa)$ as the relative permeability (μ_r) of the medium, i.e.,

$$\mu_r = 1 + \kappa.$$

Eq. 1.14.   Relative Permeability

Like κ, the relative permeability is a pure number, and in a vacuum or air is equal to 1, and therefore κ is then equal to zero.

The creation of magnetic poles at the end faces of a specimen (fig. 1.4) creates within it a reverse or demagnetizing field from the N- to the S-seeking pole. The magnitude of this effect is very much dependent on the shape of the specimen: for a long, thin needle it is negligible, while for a large, flat plate it approaches total demagnetization.

The Origins of Magnetism

At the Atomic Level

The original revelations of Ørsted and Ampère that a loop of wire carrying an electric current gives rise to a magnetic field can be transferred to the atomic scale, where each atom comprises electric charges carried by electrons (each carrying a basic unit of negative charge) and by protons (each carrying a basic unit of positive charge). The electrons move in orbits round a central nucleus, which has a balancing positive charge. A charge in orbit can be regarded as a circulating current giving rise to a magnetic field and a magnetic dipole. In addition, each electron has a spinning motion, which gives rise to a spin dipole, the moment of which is considerably greater than the orbital moment. It may be expected that

11

an interaction of these properties with an applied external magnetic field may produce some sort of magnetic response. A satisfactory explanation of the actual response of a range of materials to magnetic stimulation on the atomic scale, however, requires the application of the quantum theory of atomic structure, which first appeared early in the twentieth century. This postulates that orbiting atomic electrons exist in electronic states, or sub-shells. Each sub-shell can accept a maximum number of electrons with magnetic moments aligned in one of two (anti-parallel) directions. A filled sub-shell has an even number of electrons with equal numbers in each of the two allowed directions, giving zero resultant magnetic moment from pairing of electrons. If an odd number of electrons is present in a sub-shell, a resultant magnetic moment is created. In the absence of an external field, the spin and orbital moments are randomized by thermal motion so that there is no overall magnetization.

When a field is applied, the electron orbits tend to align with their axes parallel to the applied field direction, with rotational moments opposing the applied field to minimize the energy of the system, thus reducing the field's effect. Paired electrons, however, can make no contribution. The overall result is to create a negative magnetization, or negative magnetic susceptibility. Such materials are described as diamagnetic. Examples of diamagnetic materials are water, feldspar, quartz, and calcite. Odd electrons in sub-shells also interact with an external field through their spin moments, which align with and reinforce the field. This gives rise to a positive magnetic susceptibility, and affected material is described as paramagnetic. With significant exceptions, as examined below, many naturally occurring minerals are paramagnetic. Although diamagnetism is still present, from rotation, it is weak compared with the paramagnetic property. In fact, both are weak phenomena, with susceptibilities close to zero.

In some chemical elements, the so-called transition series, there are several unpaired electrons in overlapping shells that produce strong dipoles. In a crystal lattice with suitable dimensions, these dipoles can be linked through exchange coupling with adjacent atoms to produce magnetic domains. These multiple dipole couplings have a parallel or anti-parallel configuration. In ferromagnetic materials, the coupling is completely parallel, resulting in a very high positive magnetic susceptibility, which may persist to some extent, even in the absence of an applied field, since the strong interaction locks neighboring dipole orientations together. Iron, cobalt, and nickel are elements that exhibit ferromagnetic behavior, though only rarely in nature unless found in a pure form. However, ferromagnetic ionic compounds, such as ferric (Fe^{3+}) and ferrous (Fe^{2+}) iron, are widely present. In some minerals where large, adjacent magnetic dipoles exist, their coupling is anti-parallel, so their magnetic fields are self-canceling. If equal numbers of dipoles exhibit this, no magnetic susceptibility results , and the mineral is said to be antiferromagnetic: pure hematite (α-Fe_2O_3) is an important example of this balance. Many minerals, however, contain impurities within their crystal structures that can disrupt equality, resulting in magnetic moments. These are described as parasitic antiferromagnetics, and some deposits of hematite exhibit this property. For ferrimagnetics, though their dipoles are aligned anti-parallel, their numbers in each direction are unequal, giving rise to strong magnetic moments. This manifests as high magnetic susceptibility and spontaneous magnetization. Magnetite (Fe_3O_4), the original lodestone, is a common example of this class, and indeed most naturally occurring magnetic materials are either ferrimagnetic or parasitic antiferromagnetic (fig. 1.5). Some important properties associated with magnetism produced at the atomic level are summarized in table 1.1.

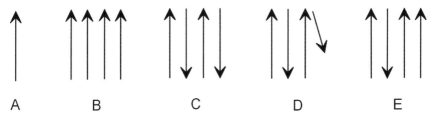

Figure 1.5. Magnetic domain orientations in different magnetic materials created by an applied field (A): ferromagnetic (B), antiferromagnetic (C), parasitic antiferrous (D), and ferrimagnetic (E).

Table 1.1. Magnetism at the Atomic Level

	Diamagnetic	Paramagnetic	Antiferromagnetic	Ferrimagnetic	Ferromagnetic
Type	anti-parallel	parallel	anti-parallel, cancellation	anti-parallel, partial cancellation	parallel
Strength	very weak	weak	zero	strong	very strong
Example	water	most minerals	hematite	magnetite	iron

In magnetic mineral structures, individual grains are occupied by magnetic domains. In some cases, only one domain is present in each, and these are described, appropriately, as single-domain (SD) forms. Other minerals have more than one domain per grain, hence are described as multi-domain (MD). In general, SD grains are smaller than MD grains.

Temperature Effects

The effect of temperature on magnetic behavior varies considerably, depending on mineral structure. For dia- and paramagnetic materials, the increasing randomization of magnetic moments from thermal agitation results in magnetic susceptibility that falls with increasing temperature. In ferro- and ferrimagnetics, the magnetic bonding within the domains is, in general, strong enough to withstand the effects of heating until a specific temperature, the Curie temperature, is reached. Then the bonds break down (within a very limited temperature range), and thereafter the material behaves as a paramagnetic. Curie temperatures are highly diagnostic of the composition of individual mineral forms, and are extensively used for identification. For the most common magnetic mineral, magnetite, the Curie temperature is 575°C. On cooling back through the Curie temperature, the behavior of magnetic minerals varies with the ambient magnetic flux density in which it is situated. All igneous minerals acquired thermoremanent magnetization (TRM) that was dependent on the magnitude and direction of the ambient field (usually the earth's) at the time of their cooling, and it resulted in a more or less permanent intensity of magnetization of a given direction. This preceded and is additional to—and often much greater than—any induced magnetization that might have subsequently occurred.

External Fields

The application of external fields to magnetic materials demonstrates the differences in magnetic structure. For dia- and paramagnetics, the induced magnetization is linear with increasing field strength, so that we can say that the magnetic susceptibility is a constant—negative for diamagnetic and positive for paramagnetic materials. The magnetization disappears when the applied field is removed. For ferromagnetics, however, this is not the case, and investigations over a range of magnetizing field strengths reveal that, dependent on the structure of the material, the application of an increasingly strong field, *H*, results in saturation magnetization intensity, M_{sat} such that further increase of the field produces no significant increase in magnetization (fig. 1.6). This can simply be thought of as all internal dipoles being aligned by the field, and no further enhancement possible. If the magnetizing field is removed, magnetization

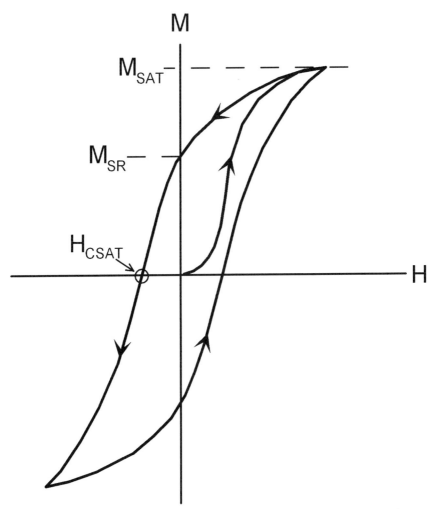

Figure 1.6. Saturation magnetization (hysteresis) loop for a ferromagnetic material. Arrows indicate the direction of magnetizing cycle: M_{SAT} is saturation magnetization, M_{SR} is saturation remanence, and H_{CSAT} is saturation coercivity.

does not fall to zero, but remains to a degree known as saturation remanence, M_{sr}, or isothermal remanent magnetization, (IRM). To remove this
remanent magnetization, a reverse field called the coercive force, or coercivity (H_{csat}), must be applied. While M_{sr} is a measure of the ease of magnetization of a specimen, H_{csat} measures its ability to retain magnetism in
the absence of an applied field. Thus soft iron, very high in remanence but
low in coercivity, is used for electromagnets, while hard steel, high in coercivity, and therefore finds use in permanent magnets. The complete cycle of saturation magnetization, demagnetization, reverse-saturation, and
return to positive saturation forms a saturation hysteresis loop (fig. 1.6),
the term hysteresis implying a lag between applied cause and resultant effect. At lower magnetic fields, when saturation is not reached (fig. 1.7), a

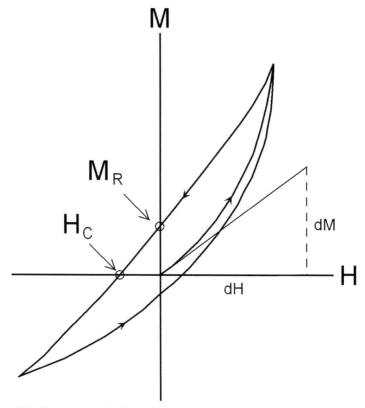

**Figure 1.7. Low magnetic field hysteresis loop: M_R is residual magnetization, H_c
coercive force, and dM/dH is initial magnetic susceptibility.**

16

hysteresis loop can still be formed by the resulting remanence (M_r) and coercivity (H_c), which are, obviously, less than the saturation values. The gradient of the hysteresis loop at any point (dM/dH) is a measure of the differential susceptibility, κ_D. At very small fields, this gradient is a measure of the zero-field susceptibility of the specimen, sometimes also referred to as low-field susceptibility, or κ_{lf}. After saturation is reached, κ_D approaches zero and the specimen behaves as a paramagnetic substance. It is worth noting here that, in the study of ferromagnetics, hysteresis loops are most readily obtained through the measurement of the resulting magnetic flux density, B, passing through a specimen, rather then the intensity of magnetization. Then the gradient of the B/H curve at any point is a measure of the relative permeability, μ_r of the specimen, from which the susceptibility is readily found.

The anomalous magnetic behavior of ferromagnetics can be ascribed to reorientation of the favored direction of magnetic domains by rotation along the direction of the applied field and by preferential growth of domains aligned toward the favorable direction. While rotation is reversible, domain-wall shift does not normally occur because it requires application of external energy via the applied field to restore initial equilibrium conditions. We will find later that all these aspects of ferromagnetic behavior are highly significant to the understanding of applications of magnetometry in the field.

Other Causes

Further forms of remanent magnetism may occur. For example, thermochemical remanent magnetization (TCRM) may occur if the effect of heating causes changes in chemical composition of the mineral. In some cases, nonthermally produced remanent magnetization can also lead to a magnetic contrast. It was found, for example (Games 1977), that unfired mud bricks may attain a weak remanent magnetization when the clay is slammed hard into a mold. Such weak shock remanence may show a slight contrast against a generally non-magnetic background and lead to the detection of relevant features. If a sample is subject to a short burst of an intense magnetic field, such as may occur in a lightning strike, isothermal remanent magnetization (IRM) may be acquired (Jones and Maki 2005).

The Geomagnetic Field

As we have seen, the presence of a magnetizing field from the earth itself has long been recognized, and, indeed, the major advances in on-site magnetic investigations of geological, environmental, and archaeological phenomena require its presence. The accepted basic model of the earth's field is that of a spinning, uniformly magnetized sphere, the magnetic properties of which can be represented by a bar magnet with its axis coinciding with the spin axis. Magnetic flux lines, emanating from within the sphere, emerge so that all matter surrounding the globe is penetrated by the flux lines (fig. 1.8). The field produced by such a magnet is said to be dipolar, although such a description strictly only applies to a magnet, the length of which is very small compared with the distance to a point of measurement. A better general definition would be bipolar. As we have previously seen, such a magnet has flux lines concentrating at the ends of the magnet, in this case at the earth's surface, at the poles described as north-seeking (N) and south-seeking (S). Since N- and S-poles attract one another, and flux lines conventionally pass from N- to S-poles, the north-seeking pole of a compass needle, pointing northward, is attracted to the south pole of the earth's magnet. The

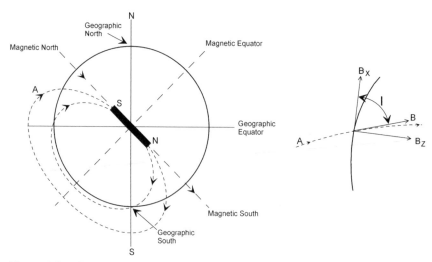

Figure 1.8. Geocentric axial dipole (GAD) model of the earth's magnetic field. *Right*, the image shows detail at the earth's surface: *I* is the angle of inclination, *B* is the earth magnetic flux density, B_x is the horizontal component, and B_z the vertical component.

earth's flux lines, entering its surface at the magnetic north pole, are thus, confusingly, converging to an S pole.

The magnetic flux density can be estimated at any point on the surface of the earth via the geocentric axial dipole (GAD) model. The true magnetic field pattern of the earth is considerably more complex than this, however, and studies of the origin of the field require a basic geodynamic effect with electric currents within the outer core of the earth, some 300 km below the surface, where molten iron circulates. Superimposed on this are variations created by perturbing, time-dependent events, and more localized ferromagnetic phenomena in the earth's crust. The resulting field is not truly dipolar, nor does its axis pass through the center of the earth. Indeed, the dipole poles are situated in the north of Canada and the south of Australia, with a long-term, time-dependent drift and strength. At these poles (and at an anomalous *second* north pole in Siberia!) there are local, vertically directed magnetic flux densities of approximately 60 microtesla (μT); the simple dipole (GAD) model (Tarling 1983, 111) would then give a horizontal equatorial flux density of 30 μT, half that at the poles.

As an approximation to the truth, the GAD model enables us to evaluate the magnitude and direction of the flux density at any point on the earth's surface. Thus intensity (B), vertical (B_Z), and horizontal (B_X) components of the field can be calculated, and the direction expressed as its angle of inclination or dip (I) to the horizontal. The angle of declination (D), defined as the angular displacement of the magnetic longitude from the geographic meridian, is zero for the GAD model but, as can be readily seen with a compass, has finite values that vary with the actual position of the magnetic poles. These parameters are routinely used in practical field observations. As can be seen in figures 1.9–1.10, based on a GAD and a Mercator projection at three latitudes, the total flux density and isoclinic lines deviate significantly from the simple model. These variations, as we shall see, are of great importance when practical magnetic field surveys are undertaken in different parts of the world.

It is clear that the earth's magnetic field varies significantly in a global sense. Generally speaking, the area of an archaeological survey is insignificant by comparison to the global rate of change. That does not mean that the variation is unimportant, however—far from it. The earth influences the shape and strength of the anomalies produced by buried features of archaeological and geological interest to our surveys.

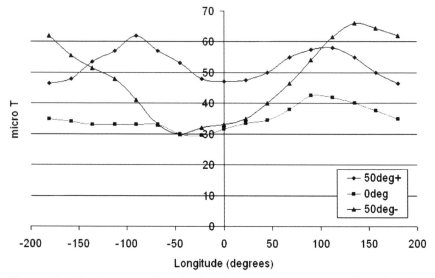

Figure 1.9. Total magnetic flux density of the earth's magnetic field at three latitudes compared with **GAD** model.

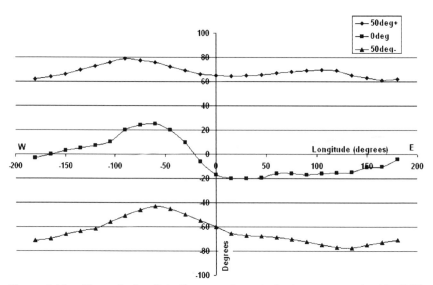

Figure 1.10. Magnetic isoclinic lines at three latitudes compared with **GAD** model.

Archaeology and Magnetism

To understand why archaeological features can be detected with magnetic prospecting methods, it is necessary to investigate how human habitation leads to the alteration of magnetic properties in the ground. Such magnetic contrast can be related either to remanent or induced magnetization.

Remanent Magnetization

The first archaeological features that were the subject of magnetometer surveys were kilns (Aitken 1958) because it was known that such structures acquire a high thermoremanent magnetization (TRM) during their firing. In natural clays rich in iron oxides, neighboring magnetic domains are almost randomly oriented; under normal conditions; the earth's magnetic field is too weak to produce significant alignment in them, and the resulting induced magnetization is relatively weak. However, when these minerals are heated above their Curie temperatures (table 1.2), the materials lose their magnetic order and become paramagnetic, so that their individual magnetic moments can readily align with the ambient magnetic flux density (B_{earth}). Once such material has cooled below its Curie temperature, magnetic order reappears, and the domains form round the newly aligned magnetic moments. It is due to this consistent alignment that the overall TRM is high, and creates a significant contrast between the heated feature, acting as a strong bar magnet aligned with the earth's field at the time of firing, and the surrounding earth (fig. 1.11). The secular change of the earth's magnetic field therefore leads to a difference between the current field direction and the TRM. As well as aiding location by magnetometer survey, this change can be used to date the time of the feature's last firing.

In addition to kilns, the effect can be seen in clay ovens, and in floors baked by hearth fires or destructive conflagrations. Fired bricks also

Table 1.2. Common Iron Oxides

Material	Composition	Oxidation	Magnetism	χ [10^{-8} m^3kg^{-1}]	Curie Temp °C
Hematite	α-Fe_2O_3	Fe^{3+}	antierro	60	675
Magnetite	Fe_3O_4	Fe^{2+}, Fe^{3+}	ferri	56,000	575
Maghemite	γ-Fe_2O_3	Fe^{3+}	ferri	30,000	575–675

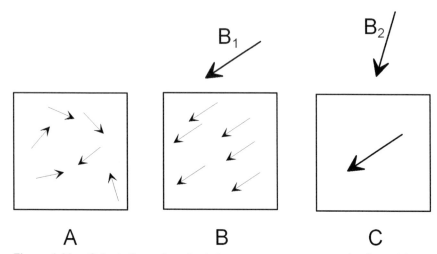

Figure 1.11. Orientation of ancient thermoremanent magnetization with respect to the earth's magnetic field: (A) before firing, (B) at firing in field B, (C) residual magnetization in later field 2.

obviously exhibit the property of TRM. However, when assembled to form a structure, the direction of magnetization in individual bricks becomes randomized, and the overall magnetization of the structure is greatly reduced (Bevan 1994, Hesse et al. 1997). In many cases, the remains of a brick building fill with brick rubble following demolition, and the contrast between structure and rubble becomes very weak. Figure 1.12 shows a magnetometer survey of a seventeenth-century house at Stowe Barton, Cornwall, where the brick foundations show as lines of reduced "noise" against the highly varying background. If a brick structure were exposed to intense heat in, say, a fire leading to its destruction, the magnetic moments of individual bricks might well become aligned with the earth's field, as with a kiln. In such a case the remains of the building would show a strong TRM value.

Induced Magnetization

Where an archaeological feature is characterized by a contrast of its magnetic susceptibility in excess of its surroundings, it will show induced magnetization in the direction of the earth's magnetic field ($M = \kappa \cdot H_{earth}$). Numerous studies (Fassbinder and Stanjek 1993; Hounslow and Chepstow-

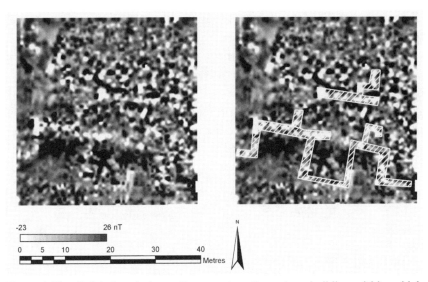

Figure 1.12. **Brick foundations of a seventeenth century building within a high magnetic noise background.** *Right,* **the positions of the walls are indicated.**

Lusty 2002; Jordanova et al. 2001; Linford 1994, 2004; Linford and Canti 2001; Maki et al. 2006, 2007; Tite 1972; Tite and Linington 1975, 1986; Tite and Mullins 1970, 1971; Weston 1995, 1996, 2002, 2004) have revealed that anthropogenically influenced topsoil has a high concentration of iron oxides that lead to enhanced magnetic susceptibility. It is necessary therefore briefly to discuss iron oxides and their relationship to human habitation. Iron oxides are the most common magnetic minerals found in the earth. They consist of assemblages of iron and oxygen ions that, depending on their particular crystal arrangement and ionic valence, show a wide rang of magnetic properties. The iron oxides most relevant for archaeological prospecting are hematite (α-Fe_2O_3), magnetite (Fe_3O_4) and maghemite (γ-Fe_2O_3) (table 1.2).

In hematite, all the iron atoms exist in the fully oxidized state (Fe^{3+}). The magnetic alignment is antiferromagnetic leading to a weak overall magnetic susceptibility. In contrast magnetite and maghemite are ferrimagnetic with approximately 1,000 times higher magnetic susceptibility. The main difference between the last two is the level of iron oxidization. In magnetite, some of the iron atoms are only partially oxidized (Fe^{2+}),

23

whereas, in maghemite, all iron is fully oxidized. It is interesting to note that, despite the same stoichiometric iron and oxygen composition (i.e., chemical makeup), hematite and maghemite have very different magnetic properties attributable to the differing arrangements of ions in their crystal lattices.

It is clear, from this brief description, that a sequence of reduction and re-oxidation can increase magnetic susceptibility by converting iron oxides into highly magnetic forms, thus:

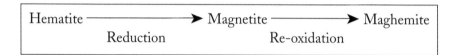

It is important that, for this enhancement, the re-oxidation of magnetite leads to maghemite. If this were not the case, then a closed reduction/re-oxidation cycle would end at the starting point without enhancing magnetic susceptibility.

One of the most common iron oxides found in soil is the weakly magnetic hematite. After initial work by Le Borgne (1955, 1960), several studies (Graham and Scollar 1976; Tite 1972; Tite and Mullins 1970; 1971) have attempted to identify its conversion into strongly magnetic oxides in concert with human habitation. Most common is the Le Borgne effect, which is linked to heating and burning, both activities long associated with human occupation. The burning of vegetation in most surface fires is sufficient to exclude oxygen, thus producing a reducing atmosphere. Under these conditions, temperatures of about 200°C are sufficient to reduce hematite in topsoil to magnetite. When the fires abate, oxygen becomes available, and in cooling the magnetite can re-oxidize as maghemite, leading to permanently enhanced magnetic susceptibility of the topsoil. Tite and Mullins (1971) replicated this process in the laboratory by covering soil samples with household flour, then heating them in reducing furnaces, followed by cooling in air. The results confirmed Le Borgne's earlier studies.

In addition to fires, human habitation is also linked with organic waste. Domestic waste heaps, or middens, and detritus are home for many microorganisms that propagate decay. These bacteria create the reducing or oxidizing conditions necessary for waste digestion. In these conditions, magnetic minerals may also be converted, leading to enhancement of magnetic susceptibility in the soil. This process is historically referred to as fermentation although, strictly speaking, methanogenesis is not necessary for such conversion to occur. A more appropriate term would be "microbially mediated" (Linford 2004).

A third pathway of enhancement is formed by certain bacteria that are able to create micron-sized magnetite crystals within their bodies by using iron oxides naturally occurring in soil (Fassbinder et al. 1990). Such magnetotactic bacteria have been found in considerable concentrations in the remains of decayed wooden posts (Fassbinder and Stanjek 1993). Although the magnetic signals produced by these bacteria are weak, they can be detected with sensitive magnetometers if the response from the surrounding soil matrix is low. For example, magnetic surveys in loess soils have revealed the remains of wooden palisades within circular neolithic enclosures in Austria (Neubauer et al. 2001).

A fourth pathway for magnetic enhancement of topsoil is by the addition of magnetic material, such as broken pottery or brick fragments (Weston 2002). Such material was often simply discarded, incorporated into middens by farmers who spread it on arable fields with other manure. Metalworking debris, for example hammer scale and slag, also becomes incorporated into soil layers to greatly increase magnetization. It is an aspect of modern life that iron and steel fragments broken or fallen from today's farm machinery can create undesirable magnetic anomalies in survey data.

Enhancement of soil magnetic susceptibility also occurs during soil formation processes (pedogenesis). Maher and Taylor (1988) reported the formation of ultrafine grained magnetite in soil despite the absence of any microorganisms. This is considered the fifth pathway of enhancement.

The first three pathways rely on the availability of organic matter, which is usually more abundant in the upper soil horizon than in the subsoil, hence creating a magnetic differentiation of topsoil and subsoil. In addition, anthropogenic input further enhances these conditions (either through fire or deposition of organic material), sometimes allowing the identification of settlement areas through magnetic susceptibility mapping, or the differentiation of buried land surfaces (e.g., covered by windblown or alluvial deposits) from the magnetic stratigraphy.

All these pathways lead to the presence of a greater abundance of magnetite, maghemite, or a combination of these two ferrimagnetic iron oxides in the surface soil or in the fill of archaeological features. Whenever a ditch or pit is filled with magnetically enhanced soil, the susceptibility contrast with the surrounding soil or sediment matrix will lead to induced magnetization, which produces a magnetic field measurable at the surface. However, archaeological features do not always show a positive magnetic contrast. Linford (1994) and Weston (2002) have tried to explain the lack of useful magnetometer signals from two British archaeological sites. Similarly, no magnetic contrast has been found on some archaeological sites under estuarine deposits in The Netherlands (Kattenberg and Aalbersberg 2004). In another extreme case study, Maki et al. (2006) investigated the origins of a negative magnetic anomaly caused by a series of archaeological hearths. If one assumes that the soils at a site are capable of producing a signal strong enough to be measured, then there are two possible explanations for the observed lack of magnetic contrast. Either contrast there has never existed, or subsequent processes have altered the iron mineralogy of the soil and thereby changed its magnetic susceptibility. Soil is a dynamic medium, and chemical changes can be caused, for example, by wetting and drying, or by persistent waterlogging. In summary, the dynamic changes and the interrelated nature of the pathways indicate that the enhancement of magnetic susceptibility is not entirely predictable. These effects are strongly site-dependent and must be investigated on an individual basis.

Fractional Conversion

As noted above, human habitation can lead to the conversion of a soil's natural iron oxides into highly magnetic forms, such as magnetite and maghemite. The resulting enhancement of overall magnetic susceptibility will therefore depend on two factors: the initial iron oxide concentration in the soil and the human impact on it. While only the latter is of concern for anthropogenic enhancement, the two factors cannot be separated easily. The absolute value of the magnetic susceptibility is therefore not very meaningful archaeologically. Where natural soil is very rich in iron oxides, it is possible to relate measurements of magnetic susceptibility to background levels taken from areas that are not anthropogenically altered. If a marked increase is detected in specific areas, these may be associated with past human activities. However, the opposite conclusion is invalid. If, for example, the measured magnetic susceptibility is low throughout, this can be the result of either a low concentration of hematite in the soil in the first place, as on chalk geology, or a lack of human activities that would have converted it into more magnetic forms.

A way to resolve such ambiguity partially is to evaluate a soil's fractional conversion (Crowther 2003, Crowther and Barker 1995). Based on the assumption that the main iron oxides in natural soil have negligible magnetic susceptibility (e.g., hematite), the actual susceptibility of a sample can be seen as an indicator of its secondary ferrimagnetic constituents (magnetite and maghemite). For a quantitative analysis, the maximum achievable magnetic susceptibility of a sample is evaluated by thoroughly exposing it to reducing conditions in a furnace (Clark 1996). The ratio of initial magnetic susceptibility to this maximum value gives the fractional conversion, and indicates how much of the weakly magnetic fraction had initially been converted. Although nonanthropogenic causes are possible, it is feasible to attribute most of this conversion to human activities. A fractional conversion of 0% hence suggests no human impact, while 100% represents intense treatment of the soil.

Archaeological Features

It is clear from the discussion in this chapter that a single measurement of magnetic susceptibility is insufficient to assess the impact of human activities, and that comparisons with background values are therefore necessary. Similarly, only the contrast of magnetic properties between natural soil and buried archaeological features allows the remote detection of the latter. For induced magnetization, this may be characterized by the magnetic susceptibility contrast

$$\Delta\chi = \chi_{feature} - \chi_{soil}$$

Eq. 1.15. Magnetic Susceptibility Contrast

A discussion of processes that may lead to the magnetic contrast in a ditch will illustrate this concept (plate 1). Prior to a more substantial habitation, the topsoil magnetic susceptibility on a site may have already increased by some process, such as surface fires. When, subsequently, a ditch was dug, it was cut through this enhanced topsoil (about 0.3 m deep) into the far less magnetic subsoil. During the occupation of the site, the excavated topsoil was mixed with other soil that became more magnetically enhanced through human activity. After the abandonment of the site, the ditch gradually filled with this enhanced material contrasting strongly with the weakly magnetic subsoil. Modern farming finally redistributed the topsoil evenly, thus sealing the magnetic susceptibility contrast underneath. In this example, the magnetic susceptibility of the archaeological feature is higher than that of the surrounding soil, leading to a positive contrast. In other situations, this contrast can be negative, the most common causes being nonmagnetic stones, such as limestone, that are buried in soil of otherwise moderate magnetic susceptibility. These can be in the form of foundations or as revetments alongside, for example, Roman roads. Other features that may show negative magnetic contrast are ditches cut into highly magnetic subsoil (e.g., possibly derived from volcanic bedrock and subsequently filled with low-magnetic alluvial soil from other areas). Where geological or environmental conditions vary considerably across a site (Schleifer et al. 2003), positive and negative contrasts may be found along the course of a single feature.

MAGNETOMETERS

The essential requirement for the detection of buried features, whether geological, environmental, or archaeological, by their magnetic properties, is the presence of a magnetizing field. Since the earth provides us with this, albeit in varying magnitudes and directions, and also some buried features possess remanent magnetization that creates magnetic flux density at the surface, all can be measured with suitable instruments. And because magnetometers require no further magnetic excitation of features or targets, the methods of detection with them are typically termed passive. Yet these instruments, now developed into sophisticated devices both accurate and robust, are rightly regarded as the workhorse of archaeological prospecting (Clark 1990) because of the speed at which they can survey large areas.

In broad terms, practical magnetometers can be classified into scalar and vector instruments. The scalar measure the total strength of an ambient magnetic field at any given point, while the vector measure a component of the field in a particular direction. Scalar magnetometers are therefore often referred to as total-field magnetometers, although intensity magnetometer might be a more accurate term. Magnetometer sensors of interest to archaeologists include the fluxgate and SQUID, both vector, and the proton, Overhauser, and alkali-vapor, all scalar.

It is worth noting that it is also possible to create an alternating magnetic field with appropriate instrumentation to excite secondary fields in the ground and to measure the resultant flux densities. Such methods are

designed to ignore the presence of the earth's field and target only localized magnetic (and electrical) properties of the earth. These electromagnetic methods are described as active, since a primary field has to be
created by an instrument in order to produce a secondary field to be measured. The metal detector is the best known example of the electromagnetic method; the susceptibility meter is a further example, and is a
well-recognized prospecting tool in its own right. It measures a magnetic
property directly, unlike a magnetometer, which monitors the flux density
created by induced magnetization.

Historically, magnetic compasses were first used in the study of the
earth's magnetic field, and developments of the compass provided the first
approach to high-sensitivity magnetometers. A magnetic needle suspended horizontally from a torsion-free fiber, such as unspun silk, will oscillate (see chapter 1) about the direction of the horizontal component of
the earth's field, B_x, with a time period

$$T = K/(B_X)^{1/2}$$

Eq. 2.1

where K is a constant depending on the magnetic moment, dimensions,
and mass of the needle.

If this instrument is calibrated in a known field to give the constant K,
then it can provide a method for measuring B_X. The accuracy of the measurement clearly depends on the measurement of T. By using a so-called
astatic pair of suspended needles with opposing poles, it is possible to produce oscillations of a very long period that can be timed with great precision, thus giving accurate flux density measurements. Variations of B_X in
the order of 0.01% are then readily measurable. A magnetic needle pivoted
so as to rotate in a vertical plane can be aligned in the direction of the
earth's field, and it will then dip along the direction of the field B to measure, directly, the angle of inclination D to the horizontal. In this way, the
elements of the earth's field at a point on its surface can be ascertained.
Such measurements, however, require laboratory conditions and are very
time-consuming. Other, more rugged, instruments employing torsion sus-
pensions have been used in global surveys of the earth's magnetic elements.
However the sensitivity achieved is at least an order of magnitude less than
is required in practical surveys with magnetometers for archaeology.

Sensor Configuration

For practical purposes, a single magnetometer sensor is limited because it measures not only the magnetic flux density of the archaeological feature but also of any underlying geological bodies, and the earth's magnetic field. The last is particularly problematic as the earth's field can change dramatically even over a short time span. The reason is solar wind, which consists of charged particles emitted at high speed by the sun. These moving charges produce their own magnetic field, which distorts the earth's field (fig. 2.1). Consequently, the field when facing the sun, during the day, is different from that at night, with a resulting diurnal variation of the measured flux density. The nuclear reactions in the sun often lead to strong and rapid ejections of charged particles that manifest themselves as magnetic storms that can be short (minutes) or long (days). To be certain that an anomalous magnetometer reading is caused by a buried feature and not just by a magnetic storm, changes of the earth's field have to be monitored and removed from data.

If only a single magnetic sensor is used, it can sometimes be assumed that during the time needed for a short traverse (e.g., ca. 20 sec for 20 m) changes of the earth's field will be small. A simple background subtraction can then be used to remove the presumed effect (Tabbagh 2003). However, the resulting data plots often still show considerable variations in the short term, and detailed data analysis can be difficult, but see chapter 5 for successful, rapid, high-data-density surveys using single-mode sensors.

For many surveyors, a more convenient and robust approach is to use two sensors (fig. 2.2). The magnetic field generated by the earth (and distorted by the solar wind) will be identical in both sensors, and therefore subtracting one reading from the other will yield only those changes that are related to features in the ground. In this differential mode (sometimes also called variometer mode), one sensor is kept at a fixed location to continuously record the earth's field. It can either be directly linked to the second roving magnetometer via a cable, and thus provide an immediate readout of the difference signal, or data can be collected at a fixed rate, and the time of measurement stored with each magnetic flux density value so that the readings from the roving magnetometer can later be corrected. In either case, it is obvious that only the roving magnetometer will measure magnetic fields that change across archaeological and broad underlying

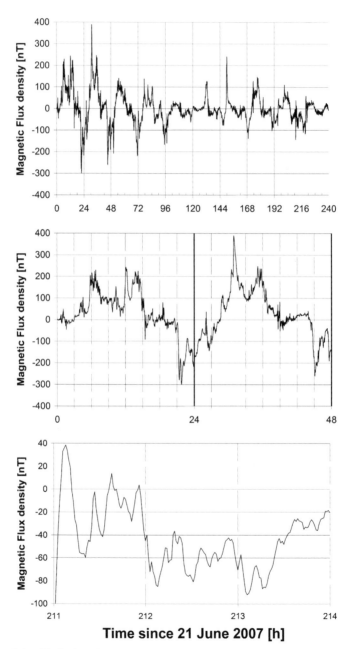

Time since 21 June 2007 [h]

Figure 2.1. Typical variations in the earth's magnetic flux density: *top*, diurnal variations in northern latitudes (10-day record); *center*, a magnetic storm (2-day record); *bottom*, micropulsations (3-hour record). Data courtesy of Geological Survey of Canada.

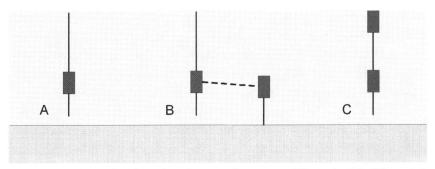

Figure 2.2. Operational modes of magnetic sensors: **(A)** single, **(B)** differential, with second sensor acting as a reference at a fixed location, and **(C)** gradiometer, with two sensors mobile, but fixed relative to one another.

geological features, or indeed as a result of nearby fences and cars. The resulting variometer data will inevitably show broad background variations superimposed on anomalies of interest.

The third option, the gradiometer, is now the most commonly used configuration with two sensors housed at a fixed distance from one another (De Vore, et al. 1994). For a vertical gradiometer, one sensor is above the other; horizontal gradiometers are also possible. The vertical sensor separation, typically 0.5 m or 1.0 m in archaeological prospection, determines the relative contributions from shallow and deep sources. At such separations, the gradiometer reading from an archaeological feature, buried typically at 1 m depth, does not approximate to a gradient reading, and gradiometer results should therefore be quoted in nT and not approximate to a true gradient, i.e. nT/m. Despite this, the two sensors are relatively close together, and as a result are affected equally by the earth's field and deep and broad geological sources. These unwanted signals are filtered out when subtracting the two readings. The gradiometer configuration therefore forms an inherent spatial high-pass filter (see also chapter 5). This effect will be most significant if the two sensors are close together. However, the closer the two sensors are together, the more of the wanted archaeological signal will be lost in the subtraction. The advantage of a short sensor separation (e.g., 0.5 m) is that unwanted local background signals are removed and the archaeological features stand out more effectively. The unwanted background may not be under the ground, and with the short separation it may be possible to work relatively close to ferrous objects (e.g., fences). However, the signal strength, and hence instrument sensitivity is reduced. In contrast, a gradiometer with a greater sensor separation (e.g., 1 m) results in a larger signal, and thus increased sensitivity to deeper features, but at the expense of eliminating less local background. It is clearly important to consider the implications of sensor separation when analyzing the resulting data.

What should be understood is that while certain types of magnetometers are often used with one particular arrangement of sensors, there is no definitive link that characterizes one sensor type with a particular mode or arrangement of sensors. Whatever sensor type is used to account for local variation of an ambient magnetic field, the way in which the sensors are arranged will be a result of personal preference, practical simplicity, and expected signal strength.

Fluxgate Magnetometers

The incentive to produce new designs of rugged, high-sensitivity magnetometers was provided by World War II, when the detection of military ordnance was paramount. The first such successful device for large-target location was the fluxgate magnetometer, which makes use of the high magnetic permeability and very low remanence of mu-metal, a ferromagnetic alloy. Saturation of this alloy occurs in fields as low as that of the earth. A fluxgate sensor comprises two short wires, or toroids, of mu-metal, which act as cores for primary coils wound closely round them so that the windings are in series, but opposing directions (fig. 2.3). Thus, in the absence of any external field, the passing of an alternating current I_p, typically of 2–5 kHz frequency, will produce equal and opposite magnetic fluxes in the two cores. A secondary coil wound round the pair therefore encompasses zero resultant magnetic flux. By the laws of electromagnetic induction, any changing flux passing through the common secondary coil induces an alternating voltage in it equal to the rate of change with time (time gradient) of the magnetic flux. Since no resultant flux is passing, there is no voltage induced in the secondary coil. As we shall see below, however, the presence of an external steady field along the cores' axes produces an output voltage proportional to the magnitude of that field.

In a typical fluxgate instrument, such as one of the Geoscan FM series, the overall dimensions of a single sensor are of the order of tens of

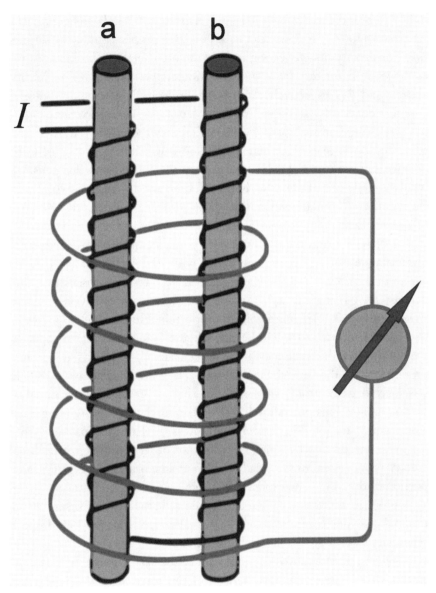

Figure 2.3.   Twin "opposing" primary-core windings, (a) and (b), of a fluxgate sensor with a common secondary winding to a detector.

millimeters and the driving alternating current, of 2 kHz frequency, is a square waveform (fig. 2.4a). Let us assume, for simplicity, that the saturation flux density magnetization loop of the mu-metal takes a linear form, as shown in figure 2.5. Here the saturation applied field and flux density are H_{sat} and B_{sat} respectively. An external uniform field H_E will produce a core flux density B_E, as shown. In the absence of such a field, the magnetization changes in each primary core will follow the square waveform until saturation is reached, when the response will be clipped equally for positive and negative fields in each primary coil, but the two flux densities will be exactly out of phase with one another, and the induced voltage in the common secondary coil will hence be zero (fig. 2.4b). In the presence of a steady field along the core axes, however, the magnetization waveform will be offset (so that the positive cycles in one primary coil are clipped at lower magnetization than the negative), while the reverse will be true for the second primary coil. Hence there will be a resultant oscillating flux density, B_S, passing through the common secondary coil with a square waveform (fig. 2.4c). Thus a voltage proportional to the rate of change of this waveform, (dB_S/dt), will be, as in figure 2.4d, a symmetrical pulsed voltage of twice the frequency of the applied signal, and with a magnitude proportional to the steady magnetizing field. In practice, the pulsed signal is subjected to electronic frequency filtering to produce a pure signal of this frequency. This procedure has the effect of virtually eliminating electronic noise generated in the coils, and gives the largest signal sensitivity for moderate driving currents. Rectification of the output gives a direct voltage proportional to the steady field, and, in a practical fluxgate, a proportion of this is returned to a feedback coil wrapped round the sensor so as to produce a further steady magnetic field that almost balances out the original one. The cores are then in a virtually zero field, resulting in high sensitivity to changes in the ambient field. The steady magnetizing field is, obviously, provided by the component of the earth's magnetic flux density, B, that is directed along the axes of the cores so that the output is highly sensitive to the sensor orientation. Indeed, if a sensor is oriented at an angle θ to the total flux direction, the component along its axis will be

$$B_\theta = B \cdot \cos\theta.$$

Eq. 2.2.

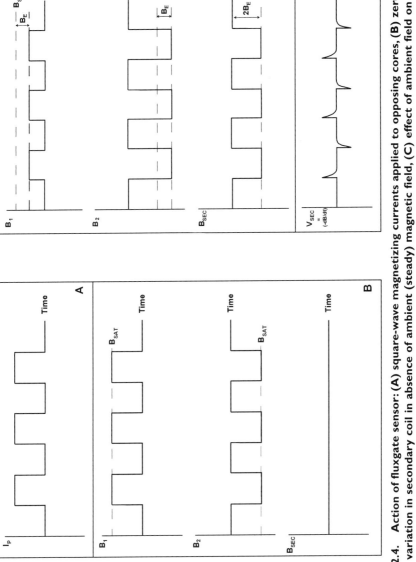

Figure 2.4. Action of fluxgate sensor: (A) square-wave magnetizing currents applied to opposing cores, (B) zero resultant flux-density variation in secondary coil in absence of ambient (steady) magnetic field, (C) effect of ambient field on core flux densities (note resultant alternating flux density in secondary coil proportional to steady field magnitude), and (D) resultant voltage output ($-dB_{sec}/dt$) from secondary coil proportional to rate of change of magnetic flux and twice the applied-field frequency.

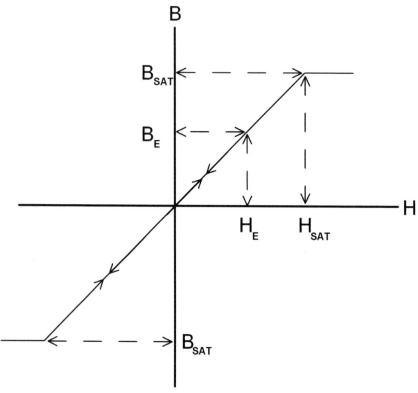

Figure 2.5. Idealized saturation magnetization loop of a fluxgate core, (H_{sat}) saturation magnetizing field, (B_{sat}) core saturation flux density, (H_E) ambient (steady) field along axis of core, and (B_E) core flux density from ambient field.

The sensitivity to angle change can be estimated by simple differentiation so that

$$dB_\theta/d\theta = (-)\, B \cdot \sin\theta.$$

Eq. 2.3.

If, typically, $B = 50{,}000$ nT and $\theta = 30°$, this gives a change of approximately 7 nT in B_θ for a change in θ of one minute of arc. As is evident from this equation, the sensitivity can be reduced if θ is close to zero and the sensor axis and flux direction are closely coincident. For a vertical fluxgate sensor, this would be near the magnetic pole. During early experiments with fluxgate magnetometers, the sensors were flown attached to aircraft, and high flux densities measured when it was adequate to approximate to such an alignment. For terrestrial use, however, where high portability and the detection of small field changes are required, it is usual practice to orient the fluxgate so that its axis is vertical, and thus measures B_Z. For an inclination of 60°, i.e., at moderate latitudes, the angle θ between fluxgate axis and field direction would be 30°, as in the above example, and it is clear that adequate steps must be taken to maintain the orientation of the fluxgate and minimize any tilt from the vertical. This is achieved by constructing the instrument in a gradiometer form (fig. 2.6),

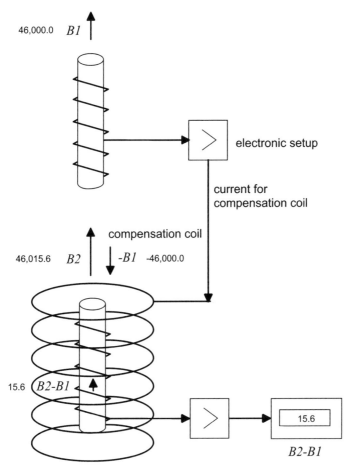

Figure 2.6. Fluxgate gradiometer principle. Output from upper sensor feeds in opposition to lower sensor to provide output B2 − B1. In the absence of a local magnetic anomaly, the electronics are adjusted to set the resultant to zero.

which also eliminates the influence of the unwanted variations noted above. Two identical fluxgate sensors, (1) and (2), are mounted in opposition, accurately parallel to one another on a rigid, portable staff, with one at a fixed distance above the other. The output of the instrument is then a voltage, which is proportional to

$$B_{lower} - B_{upper,}$$
Eq. 2.4.

where B_{lower} and B_{upper} are the flux densities in the lower and upper sensors respectively. We will show in chapter 3 that a gradiometer can eliminate the effect of the earth's field. In a fluxgate gradiometer, the tilt-related flux density in equation 2.3 is then only the one created by the buried feature. For example, for a 10 nT anomaly, even a 1-degree tilt only produces a change of 0.1 nT. The tilt problem is now greatly reduced and, by appropriate angular adjustment of the sensors, a high degree of parallel alignment can be achieved. It is evident that alignment errors resulting from the setup procedure are a problem, and these are inherent in all fluxgate instruments. However, these can be kept to a minimum by robust instrument design and good field practice (see chapter 4).

The first fluxgate gradiometer to be developed specifically for archaeological prospecting appeared in 1964 as a result of the pioneering work of John Alldred at the Oxford Laboratory for Archaeology and the History of Art. After Alldred's (1964) initial foray into this area, Frank Philpot (1973) built an instrument that was based on a smaller separation between the sensors; this 0.5 m distance became the standard separation in instruments built for archaeological purposes for about 30 years. Philpot was also influential, along with the Ancient Monuments Laboratory in England, in designing a continuous recording system using a chart recorder. As a result of this development, a skilled operator could survey and produce a visual image of data within the same working day. However, a major leap forward in the practical use of fluxgates came as a result of integrating a data logger with the FM series of instruments (www.geoscan-research.co.uk). We will examine the performance of the fluxgate gradiometer in greater detail later, when we look at potential targets that lie in the near surface.

Fluxgate Magnetometer Summary

- Vector (usually vertical component)

- Usually gradiometer mode

- Sensitivity ca. 0.1 nT

- Heading Errors

- Low power and weight

- Significant setup for some types

- Significant drift (thermal, mechanical) in older models

- Used in areas of high gradient

- Relatively cheap

Proton Free-Precession Magnetometers

Hydrogen is the simplest of atoms, comprising a nucleus of one fundamental particle, the proton, carrying a positive unit of atomic charge,
around which circulates an electron carrying an equal and opposite charge.
The proton spins about a specific axis at a rate defined by its quantum
state, and as such behaves like a magnetic dipole of atomic size with its
axis along the axis of spin. It will therefore be influenced by any applied
magnetic field in such a way as to align with the direction of the flux density. However, as in the case of a compass needle, which oscillates in two
dimensions about the final rest position with a definite frequency, the proton gyrates or precesses in three dimensions with a precession frequency
f_P that is directly proportional to the magnetic flux density (fig. 2.7). The
constant of proportionality depends only on the charge and spin properties of the proton, and is known as the gyromagnetic ratio. Its value is
known with great accuracy. Indeed, we can state that

$$f_P = B \times 0.042577 \text{ Hz/nT},$$

Eq. 2.5.

where B is the intensity of the magnetic flux density, around which the
proton precesses. In a volume of hydrogenous fluid, such as water or

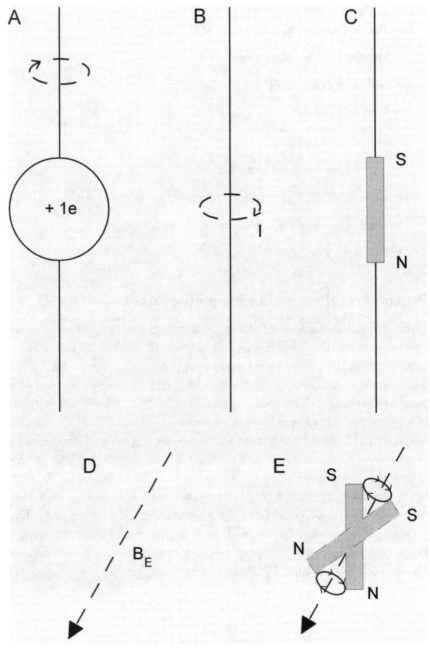

Figure 2.7. Free precession proton principle: (A) spinning proton of +1 electron charge, (B) equivalent current loop, (C) equivalent magnetic dipole, (D) ambient magnetic flux density, and (E) precessing dipole within the ambient field.

methanol, there is a myriad of such atoms attempting precession around the direction of the ambient magnetic field. However, because of thermal agitation, these attempts are constantly randomized, so that no resultant magnetic effect can be observed. In the proton free-precession magnetometer, a measure of control is exerted on the hydrogen atoms, contained in a "bottle" of appropriate fluid, by applying a polarizing magnetic field that is very much stronger than that of the earth and at an angle to it. This is achieved by means of a coil closely wrapped around the bottle. Electric current in the coil creates a field that polarizes the proton dipoles along its direction, overcoming, to an extent, the permanent thermal randomizing influence. Thus a significant proportion of the proton population is aligned in the direction of the polarizing field. If the polarizing field is now removed, this population swings back to precess around the direction of the ambient field, initially in a coherent fashion, so that a measurable gyrating magnetic field, of frequency f_p, is created. The oscillating flux density of this field will produce an induced voltage in the now "quiet" polarizing coil, where its frequency can be measured using appropriate electronics.

Assuming a magnetic flux density of the order of 50,000 nT for the earth's field, a precession frequency of approximately 2 kHz is created. This is readily measured to six-figure accuracy using modern electronics, so that, using equation 2.5, the earth's total magnetic flux density can be measured to a fraction of a nanotesla. Unfortunately only a limited time is available for such a measurement, because the protons in the hydrogenous fluid readily relax to their random gyrations as a result of thermal agitation (fig. 2.8). This relaxation time is approximately three

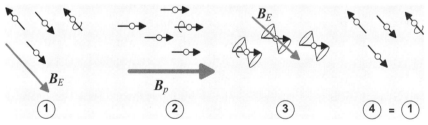

Figure 2.8. Free-precession proton magnetometer principle: (1) spin axes aligned to ambient flux density B_E, (2) polarization in presence of an applied flux density B_p, (3) precession around ambient flux density direction on removal of B_p at a frequency defined by B_E, and (4) return to initial alignment.

seconds for impurity-free water; the presence of impurities such as salts significantly reduces the relaxation time. Cycling of polarization and relaxation times can be repeated at intervals of up to six seconds, so that reliable measurements of B can be made.

In a typical, practical instrument, a polythene bottle containing about 200 ml of ion-free water (or methanol for use in temperatures below 0°C) is wound with approximately 1,500 turns of copper wire. A polarizing current of 1 ampere is passed for up to 3 seconds, producing a flux density, B_p, of about 20 mT, which is nearly three orders of magnitude greater than that of the earth. The bottle axis is normally aligned roughly magnetic east-west so as to obtain the optimum precession performance when the polarizing field is switched off (fig. 2.9). The precession frequency is measured with an appropriate phase-sensitive system and converted to flux.

A significant advantage of the proton magnetometer over the fluxgate is that it measures the intensity of the ambient magnetic flux density and as a result is insensitive to orientation problems. However, it does not record continuously, requiring a measuring time of four to six seconds, and problems are encountered when the sensor is in a high magnetic field gradient, such as might be found near a large ferrous anomaly. Under these conditions, the precession frequencies at each end of the bottle may be significantly different, leading to a more rapid loss of coherence and a relaxation time down to less than one second, which is inadequate for reliable readings. As with a single fluxgate sensor, the occurrence of regional variations of the earth's field from diurnal changes and nearby electrical installations produce an undesirable background. Use of a fixed station reference magnetometer to record these changes in the variometer mode, or a gradiometer arrangement of two sensors on a vertical staff, can greatly reduce this problem. In either case, there is no requirement for accurate alignment of the bottles with respect to one another, since the two sensors measure the intensity of the flux density. However, if the bottles are aligned with the earth's magnetic field, the polarizing field will be unable to initiate a precession, producing no readout (a dead zone), and therefore has to be avoided. In the gradiometer configuration, a long staff can be used to keep the sensors a set distance apart, and the only limitation is the portability of the system. For shallow archaeological features, the remote sensor stays virtually uninfluenced by the target, acts as a monitor of the earth's field, and can be used to strip away unwanted deeper signals.

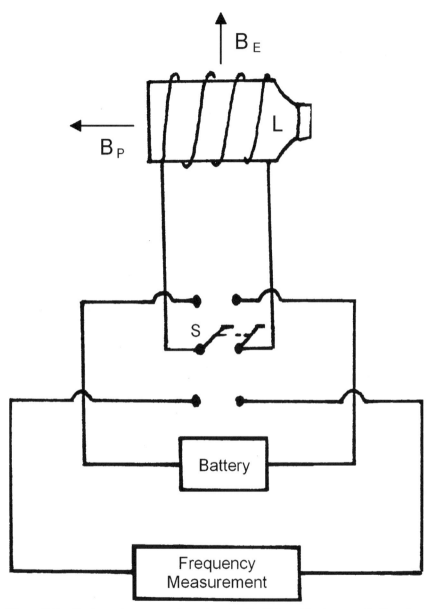

Figure 2.9. Basic proton magnetometer circuit: (B_E) ambient magnetic flux density, (L) sensor bottle containing hydrogenous liquid, (B_P) polarizing flux density, and (S) two-way switch.

The history of the application of the proton magnetometer for archaeological purposes is slightly longer than that of the fluxgate gradiometer. In 1957, the Canadian scholar John Belshe used a proton magnetometer, designed at Cambridge University for marine use, to chart the change in magnetic response over an experimental kiln (Scollar et al. 1990; p. 514). However, a pivotal moment came after a lecture on magnetic dating by Belshe at Oxford. Inspired by his presentation, Graham Webster, a Roman-period archaeologist, wondered if a proton magnetometer could be built to detect kilns that were believed to be buried in the footprint of a proposed road-widening. The two major players, Martin Aitkin and the late Edward Hall, of the same Oxford laboratory as Alldred (noted above), had heard that the details of a proton magnetometer had just been published by Waters and Francis (1958). However, the road scheme was only a matter of eight weeks away from construction, and that window of opportunity had to include designing, building, and testing the system, and then surveying the road corridor. Aitken (1958, 1986) has documented this short but significant period in magnetometer history. Among buried pipes and bedsteads, they finally located a kiln with their proton magnetometer, as well as filled-in rubbish pits. Initially, detection of the pits was not particularly of note to Aitken, who saw their response as noise that was cluttering the data from the kiln. The archaeologists involved in the project thought differently, and the significance of the results from the proton magnetometer was accepted. This instrument was clearly capable of measuring very slight perturbations in the earth's magnetic field, and therefore a great variety of marginally magnetic targets could be identified.

Proton Free-Precession Magnetometer Summary

- Scalar (total-field)

- Single, differential, or gradiometer mode

- Sensitivity ca. 0.1–0.5 nT

- Effectively no heading errors

- High power consumption

- Dead zone

- Simple system, easy to use
- Slow data acquisition
- Radio frequency interference
- Problems in high magnetic gradients
- Low cost

Electron Spin-Resonance (Overhauser) Magnetometers

This type of magnetometer was originally developed in France by a group led by Antoine Salvi (1970), and a commercial version was subsequently produced by GEM Systems Advanced Magnetics. Also called an Overhauser magnetometer, it relies on, and gets this name from, the enhancement of the polarization effects seen in a proton magnetometer by the addition of a second liquid containing free unpaired electrons, or free radicals, in its composition. In place of a steady polarizing field, a very high-frequency (MHz) but weak magnetic field is used to polarize the free electrons, which then force the protons into alignment (the Overhauser effect). This combination effectively raises the ease of polarization by several thousand times over that of the proton instrument, and as a result very low power consumption is required. The precession frequency signal is picked up by a sensor coil, and after amplification a fraction of it is returned through a second coil to give positive feedback and create resonance of the output signal to enhance the performance of the instrument. Thus the Overhauser magnetometer output is virtually continuous, typically five samples per second, with a routine sensitivity better than 0.05 nT (Hrvoic et al., 2003). The instrument measures the intensity of magnetic flux density, and is therefore omnidirectional. Its low power requirement gives it much greater portability than the proton magnetometer, which it will, its manufacturers claim, supersede. At present, however, this instrument is not often used for archaeological purposes.

Overhauser Magnetometer Summary
- Scalar (total-field)
- Single, differential, or gradiometer mode

- Sensitivity ca. 0.05 nT

- Virtually no heading error

- Little power demand, light and compact

- No warm-up time

- Good gradient tolerance

- Fast sampling (many per second)

- Relatively costly

Alkali-Vapor (Optically Pumped) Magnetometers

These magnetometers are based on the properties of the members of the alkali group of atoms (sodium, potassium, rubidium, and cesium), which each carry one valence electron per atom. As noted in chapter 1, an electron in orbit can, by quantum rules, only occupy one energy state, E, at a time, within which its spin and orbital motion can be defined as clockwise or anticlockwise, with resultant directional magnetic moments. An electron can be raised to a specific excited state, E_2, from a ground state, E_1 only by the application of a discrete excitation energy, and will subsequently lose that energy in returning to the ground state by emission of electromagnetic radiation of frequency ν, such that

$$h \cdot \nu = (E_2 - E_1),$$
Eq. 2.6.

where h is a fundamental constant called Planck's constant, with a value of $6.6260693 \times 10^{-34}$ Js, or joule-second. This is an entirely reversible process. The application of radiation of exactly the same frequency ν will raise an electron from the ground state to state E_2. If alkali atoms in vapor form are rendered luminous, the emitted radiation contains discrete, characteristic line spectra created by the transition of excited electrons to lower energy states. When a magnetic field is applied to produce a flux density B in the luminous vapor, the observed lines are seen to split into a fine structure of multiple lines. The separation of these lines, $\Delta\nu$ (or ΔE), is very precisely proportional to B. This splitting of energy levels in the magnetic field can be ascribed to the interaction of the magnetic moments of individual electrons with the applied flux density and is a quantum phenomenon. Precession, at a frequency (called the Larmor precession frequency) defined by the chemistry of the atom, interacts with the electron energy state so as to change it by the precise amount of the precession energy. Consequently, an original single energy level is split into two (e.g., E_{1+} and E_{1-}), corresponding to the difference in precession energy and, hence, B. This is known as the Zeeman effect. Furthermore, when this Zeeman radiation is examined, it is found to be circularly polarized; that is, the basic electric vector identified with electromagnetic radiation is rotating in a plane perpendicular to the direction of propagated radiation. The direction of this rotation is either clockwise or anticlockwise, depending on the Zeeman line examined, and can be ascribed to the direction taken in the precession of the electron in the magnetic field. If the electron assembly is irradiated with circularly polarized light with the correct frequency for excitation to energy state E_2, the electrons with the same precession direction as that of the light polarization are favored for excitation. Thus level E_{1-} will absorb radiation and lose electrons to E_2 (fig. 2.10a). The electrons are unstable in this excited state, and decay spontaneously to both E_{1+} and E_{1-} because their precession directions have been randomized. Eventually this selective depletion and re-population leads to level E_{1+} being fully populated, and E_{1-} depleted, so that no further absorption of the incident radiation is possible. The ingenuity of the alkali-vapor magnetometer lies in the restoration of equality by a depolarizing process that provides the precise energy, ΔE, to transfer electrons from state E_{1+} to E_{1-} in the form of radio-frequency radiation of frequency ν_R, such that

$$h \cdot \nu_R = \Delta E.$$
Eq. 2.7.

Measurement of ν_R gives ΔE, and hence a precise value of B.

In practice, an alkali-vapor magnetometer (fig. 2.11) comprises a glass cell, some 30 mm in length and diameter, containing the alkali metal in vapor form. To maintain the vapor state, the cell temperature is raised to about 50°C. The cell is illuminated by light from a discharge lamp containing the same alkali-vapor raised to incandescence using radio-frequency induction

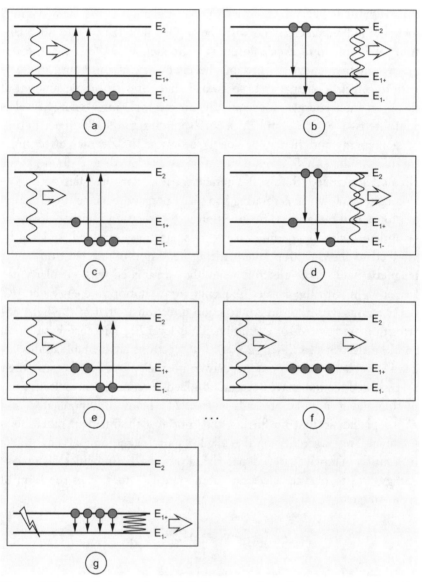

Figure 2.10. Optical pumping: (a) Before pumping, all atoms are in their E_{1-} ground state; on applying a beam of light of appropriate energy and polarization, the atoms absorb radiation and rise to level E_2; (b) these atoms are unstable and drop back in equal numbers into E_{1+} and E_{1-}; (c to f) this continues until all atoms are in state E_{1+} and E_{1-} is totally depleted; and all light passes through (g). By applying a radio-frequency magnetic field, all atoms simultaneously drop from E_{1+} to their ground state E_{1-} so that pumping can resume again at (a). Through appropriate feedback electronics, the frequency of the cycled pumping (a to g) becomes $\nu_R = E_{1+} - E_{1-}$ and is hence a measure of the magnetic field.

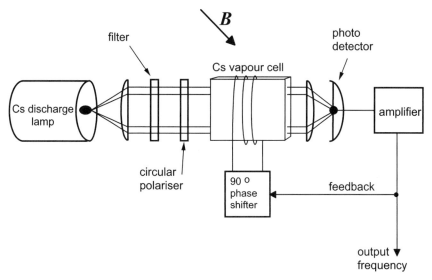

Figure 2.11. Basic cesium-vapor magnetometer circuit. The photocell output at the frequency v_R is partially positively fed back in the appropriate phase to produce an oscillating magnetic field along the axis of the cell that resonates with the pumping frequency to enhance its effect.

heating. This emits the required radiation for excitation of the sample cell vapor, together with much unwanted radiation that is removed by an appropriate filter after collimation. A parallel beam of selected light is then passed through a polarizing filter to produce circularly polarized light, then through the cell, and finally to fall on an appropriate photocell detector. When transmitted light is absorbed by the vapor in the cell, there is a reduction in the light intensity falling on the detector. Electrons falling back from E_2 to the ground state again emit radiation, but it is omnidirectional, and therefore does not contribute greatly to the photocell illumination. When level E_{1-} is totally depleted, absorption ceases and the full light transmission is restored to the photocell (fig. 2.10f). The necessary energy to repopulate E_{1-} is provided via a coil wound around the sample cell such that the magnetic flux it produces is parallel to the cell axis, which in turn is orientated approximately at right angles to the earth's magnetic field. The coil is fed by an alternating current, the radio frequency of which is swept until it coincides with v_E when energy is available to lower electrons to the E_{1-} level and to resume absorption of light.

Thus a flickering signal falls on the photocell at a frequency equal to that of the drive radio frequency and the precession frequency of the particular alkali vapor in the cell. For cesium, this is 3.49872 Hz/nT; for potassium, it is twice that value. Thus, for a typical earth flux density of 50,000 nT, the precession frequency for cesium is approximately 175 kHz, a value readily measured with great accuracy by electronic means. In practice, the small oscillating signal of about 0.1 mV magnitude picked up by the photocell is in part returned to the radio-frequency coil source with positive feedback, so that the coil signal is enhanced and maintained at the correct frequency effectively through a resonance process.

The alkali-vapor magnetometer responds virtually continuously to the magnitude of the flux density passing through the sample cell, but a dead zone occurs, which, for a cesium sensor, results in a reliable signal only when the flux lines are directed between 10 and 80 degrees to the sensor axis. The reason for this dead zone is related to the multiplicity of Zeeman line-splitting, causing a shift of the optimum spectral peak when the magnetic flux direction changes. In practice, this results in a heading error as the sensor is rotated. With appropriate instrument design, however, this can be reduced to sub-nT levels. In some practical systems, the operator moves around the sensors, thereby the sensors are always oriented the same, reducing heading errors to a minimum. The overall sensitivity achievable with a cesium magnetometer is reported as 0.05 nT with a sample rate of 10 per second. For potassium-vapor, the Zeeman splitting multiplicity is small, so that heading errors are greatly reduced. It is reported that a potassium-based system has a sensitivity of 0.009 nT at the 10-per-second sample rate (Hrvoic et al. 2003), although that claim has been contested by others (Johnson and Smith 2003). Irrespective of these counterclaims, the overall sensitivity of alkali-vapor magnetometers has lowered the detection level by an order of magnitude relative to the flux-gate and proton instruments.

The first documented use of an alkali-vapor detection device for archaeological prospecting was by the American scholar Elizabeth Ralph. She used a system, produced by the Varian company, at the site of the Greek colony of Sybaris in southern Italy. Although she experimented with a proton magnetometer, she was drawn to a rubidium-vapor device, which she used both in single and differential modes, because its sensitivity was much greater. At Sybaris, many of the features were beneath several meters of alluvium, and the 100 times greater sensitivity, as low as 0.01 nT, was ideal for the exploration she was undertaking (Ralph 1964). In a laborious campaign, she and her team collected about 400,000 hand-logged readings (Bevan 1995). During the last decade, the practical use of these instruments has forged ahead, and many novel systems have progressed beyond the experimental. In fact, in certain parts of the world, cesium-vapor magnetometers have become the standard device for archaeological prospecting. The instruments are often arranged as multi-sensor arrays on human- or machine-pulled carts, where they are used in all operational modes.

Alkali-Vapor Magnetometer Summary

- Scalar (total-field)

- Single, differential, or gradiometer mode

- Sensitivity ca. 0.01 nT

- Small heading errors

- Large power demand, large batteries

- Easy setup

- Dead zone

- Relatively expensive

Cryogenic SQUID magnetometers

The use of cryogenic SQUID (superconducting quantum interference device) magnetometers in the laboratory for measurement of very weak levels of magnetization has led to the development of new, portable systems for field use in geophysical prospection (Chwala et al. 2001; Schultze et al. 2007). The laboratory-based systems are, typically, capable of measuring flux densities of the order of 1 fT (10^{-6} nT). However, the SQUID system, like the fluxgate, requires magnetic flux lines to be accurately aligned with sensor axes, and is therefore used to measure a vector component; usually the vertical component of changes in the earth's magnetic flux density.

SQUID magnetometers rely on the property of metals whose electrical conductivity increases as temperature falls and that at low temperatures (e.g., 10K = −263°C) become superconducting, offering no resistance to electric current. This property has enormous industrial potential, and has been an area of intense research for a number of years. Its relevance to the detection of low magnetic flux densities, however, lies in the fact that, if a magnetic flux, Φ_L, is passed normally through the plane of superconducting loop, a current, I_S, is started in the loop to create a magnetic field that exactly opposes the applied flux. This is known as the Meißner-Ochsenfeld effect. However, because of the quantum nature of the superconducting phenomenon, the relationship between increasing flux density and the current is not continuously linear. At specific values of the applied flux, the flux lines are able to penetrate the loop, and the current I_S drops. Superconductivity quantum theory shows that, for the magnetic flux Φ_L to penetrate the loop, it must be a multiple integer, n, of the magnetic flux quantum number Φ_O, where $\Phi_O = 2.10^{-15}$ Wb, that is when

$$\Phi_L = n.\Phi.$$

Eq. 2.8.

Thus an increasing magnetic flux will generate a sawtooth relationship between Φ_L (or B_L) and I_S (fig. 2.12).

The Meißner-Ochsenfeld effect can be greatly enhanced if the loop is broken, then filled by an extremely thin insulating layer, a so-called Josephson junction. This was predicted in 1962 by the British physicist Brian D. Josephson, who hypothesized that, by tunneling of electron pairs across such a junction, a greatly increased superconducting current would be generated. A Josephson junction strongly enhances the sensitivity of a SQUID device to small magnetic flux changes. Beneficially, by the use of two such junctions, the sawtooth waveform can be made symmetrical, so as to create an alternating current. If an alternating flux of appropriate magnitude, rather than a steady flux, is applied perpendicular to the loop, the circulating current is clipped when the flux reaches cutoff points during the flux oscillations. These rapid current changes can be detected in a pickup coil as voltage pulses with an appropriate frequency pattern depending on the alternating drive frequency. If now a steady flux is super-

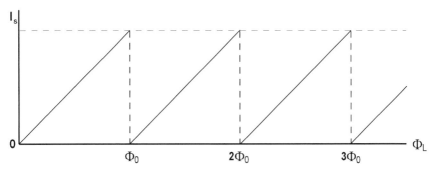

Figure 2.12.   Effect of increasing magnetic flux Φ_L on current I_s circulating in a su-perconducting loop. The current is clipped at multiples of the magnetic flux quan-tum number, Φ_0, to produce a sawtooth current variation. Application of an alternating flux produces a time-dependent circulating current, the phase of which is proportional to the magnitude of a superimposed steady magnetic field.

imposed, these pulses are displaced in time by an amount proportional to the magnitude of the steady flux, thus producing a phase shift in the fre-quency spectrum that is used to measure the magnitude of the steady flux density. The application of a superimposed alternating field to generate a response proportional to the steady field is very similar to the action of the fluxgate sensor. By using a radio-frequency current of about 20 MHz to drive the loop, changes of magnetic flux density of the order of 10^{-6} nT are achievable. Such sensitivities are unrealistic in the much "noisier" en-vironments found in fieldwork, but the construction of a gradiometer with sensors separated by a few centimeters provides a measurable magnetic-flux gradient down to 1 pT/cm, some 200 times smaller than the value that can be seen with an alkali-vapor gradiometer. Clearly, this is a very good approximation of the gradient of the earth's magnetic field, as op-posed to the very coarse representation by 1.0 m, or even 0.5 m separation, fluxgate systems. However, given the high sensitivity, it is relevant that some influence of the static ambient field still remains, and additional measurements of the field vector therefore have to be used to remove the effect of small orientation changes. As with the fluxgate, the flux density to be measured must be perpendicular to the sensor loops.

The main limitation in the use of SQUID magnetometers for geo-physical work lies in the need to maintain sensors at liquid-helium tem-peratures (approximately 4°K) for the duration of a survey that might cover several acres or hectares. Using a cryostat to maintain this temperature is cumbersome, and the cost of liquid helium is high and an ongoing cost for each survey. Still, the resolution obtained, relative to that of other magnetometers, can be impressive. More recent developments in sensor materials have seen the element niobium for liquid-helium operation (conventionally referred to as low-temperature SQUID, or LTS) replaced by complex compounds, typically yttrium-barium-copper oxides, for which superconductivity occurs at liquid-nitrogen temperatures (77°K) (Bednorz and Müller 1986). With the lower cost of liquid nitrogen, held in a simpler cryostat, such a high-temperature SQUID (or HTS) would appear to show great promise. Tests (Chwala et al. 2003) suggest that significant improvements in instrumentation are needed before the HTS technique becomes routine, but the likelihood of true tri-directional-HTS systems is clearly in sight. In addition to the high sensitivity of SQUID magnetometers, their main advantage is the very high speed with which such accurate measurements can be recorded. Sampling rates of several thousand measurements per second enable magnetometers to be mounted on nonmagnetic carts towed at speeds up to 45 km/h (28 mph).

SQUID Magnetometer Summary

- Vector (mainly vertical, horizontal possible)

- Usually gradiometer mode, true gradient

- Sensitivity ca. 0.00001 nT

- Heading Errors

- Low power, but low temperature

- Significant processing, very fast data collection

- No dead zones

- Used in high-gradient areas

- Built as gradiometers, filter geological variations

- Initial and continuing costs, very expensive

MAGNETIC ANOMALIES
AND SOURCES OF NOISE

W e have seen how naturally occurring minerals beneath the earth's surface may be modified anthropogenically to become magnetized, and how the presence of this magnetic enhancement may be detected using magnetometers designed to measure extremely small deviations in the earth's magnetic field. Clearly, if a magnetic archaeological feature has a magnetic susceptibility that differs from that of the surrounding soil, there will be a magnetic contrast. We have already discussed the possible contrast in magnetic susceptibility leading to an induced magnetization of $M_i = \Delta\kappa \cdot H_{earth}$, where $\Delta\kappa$ is the difference in magnetic susceptibility. Similarly, a contrast in remanent magnetization M_r may have been produced by past human activity. The firing of a kiln may have created a considerable thermoremanent magnetization that stands out from the surrounding soil through its magnitude and orientation in the ancient direction of the earth's magnetic field. In some cases, a feature may possess both induced and remanent magnetism leading to an overall magnetization $M = M_i + M_r$ where both contributions are added vectorially. The so-called Königsberger ratio, $Q = M_r / M_i$ is a measure of their relative strengths. The magnitude of Q is of particular significance when anomalies due to igneous geological inclusions are encountered on an archaeological site. The magnetic flux density, B, created by magnetization outside the causative feature is superimposed on the earth's magnetic field so that a magnetometer measures the total flux density

$$B_{tot} = B_{earth} + B$$
Eq. 3.1.

where B is called the magnetic anomaly due to the feature.

The Magnetic Dipole

The simplest object for which the magnetic flux density can easily be calculated is a bar magnet of length L with north- and south-seeking poles of strength $\pm\, p$, an object to which we will refer to as a bipole. The magnetic moment of the bipole is $m = L \cdot p$, and is a vector oriented along the axis of the magnet pointing from south to north. As discussed in chapter 1, these poles are purely fictitious since no magnetic monopoles have ever been found. If a bipole is cut in half, two new bipoles are created. It is possible to calculate the magnetic flux density produced by a bipole by the addition of vectors. However, a special and simpler case can be considered where the distance r from the point of observation to the center of the magnet is considerably greater than the magnet's length L. We will describe this as a magnetic dipole. To express its field in a reasonably simple way, it is useful to introduce polar rather than Cartesian coordinates. The magnetic flux density is now described in terms of its radial component B_r and the perpendicular (tangential) component B_θ (fig. 3.1). Each measurement position can be characterized by its distance r to the dipole's center and the angle θ between this direction and the alignment of the dipole. In this case, the respective coordinate values of the magnetic flux density become

$$B_r = \mu_0\, 2m \cdot \cos\theta/r^3,$$

and

$$B_\theta = \mu_0\, m \cdot \sin\theta/r^3,$$

and the intensity of the flux density B is given by

$$B = \mu_0 m \sqrt{1 + 3\cos^2\theta}/r^3.$$
Eq. 3.2.

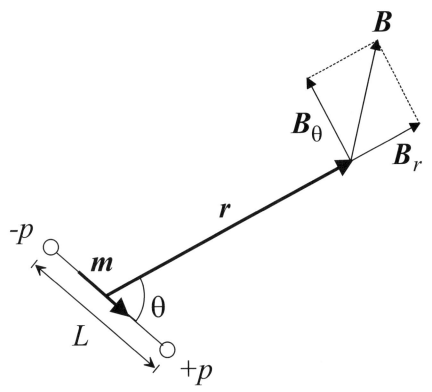

Figure 3.1. Magnetic dipole field components. Here the dipole length L is assumed to be negligible compared with the distance r to the measuring point; B_r is the radial component, B_θ the tangential component, and B the resultant flux density.

To visualize the variation of the field as an observer moves at a fixed distance r, around the dipole, a polar diagram can be used. The radial flux density component varies as $2\cos\theta$, while the angular component varies as $\sin\theta$. This is demonstrated by the diagram of "clubs" seen in figure 3.2. For the special cases in the direction of m and perpendicular to it, the magnetic flux components and intensity are therefore given as

$$\theta = 0°: B_r = \mu_0 2m/r^3, B_\theta = 0, B = \mu_0 2m/r^3,$$

and

$$\theta = 90°: B_r = 0, B_\theta = \mu_0 m/r^3, B = \mu_0 m/r^3.$$

Eqs. 3.3.

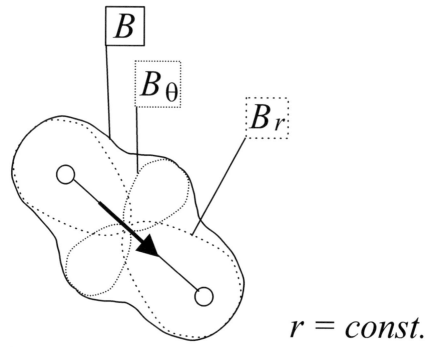

$$r = const.$$

Figure 3.2. Magnetic dipole field polar diagram of B_r and B_θ at a constant distance r.

Three important points arise from equations 3.3. Firstly, the dipole field equations can be applied as a simplification of the magnetic field of the earth (simple dipole model), as described in chapter 1. The angle θ then identifies with the co-latitude of a point on the earth's surface. Secondly, the relationship between magnetic flux density and distance from the dipole can be approximated as $B \propto 1/r^3$, which means that the field decreases very strongly as one moves away from a dipole. In other words, a small change in magnetometer sensor height above the ground over a shallowly buried dipole will alter the recorded field value considerably. Thirdly, the magnetic flux density depends on the magnetization of the buried feature. As a result, it is impossible to use the value of measured flux density to derive the depth of the feature unless a reasonable assumption can be made about the magnetic moment, or, for induced magnetization, the magnetic susceptibility and volume of the feature.

In archaeological prospecting, a sensor is traversed over an archaeological feature in the earth's field so that it remains at the same elevation above the ground. In this case, the distance r between the sensor and the dipole is given by Pythagoras's Equation $r^2 = x^2 + z^2$, where x is the lateral displacement from a point directly above the center of the dipole, moving, say, from south to north, and z is the height of the sensor above the dipole.

This flux density is characterized by intensity and orientation, and is therefore described as a vector; equation 3.1 is only correct when written in its vectorial form. Simplistically, B_{tot}, the resultant, is the diagonal in a parallelogram of vectors, as illustrated in figure 1.2, and only when B_{earth} and B are parallel to one another is B_{tot} the algebraic sum of the two component vectors. For induced magnetization, this evidently occurs along the axis of magnetization (fig. 3.3). Any other point in the local field created by the anomaly may, depending on the direction of B, be less than B_{earth}. In

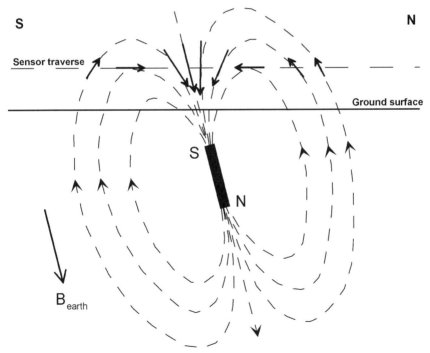

Figure 3.3. Variation of the anomaly total flux density **B** in a south-north traverse of a sensor over a magnetic bipole in the northern hemisphere. The measured flux density B_{tot} will be the vector combination of B and B_{earth}. The lengths of the arrows on the sensor traverse indicate the magnitude of **B**.

the case of remanent magnetization, the total flux density will be the vector sum of the earth's field and an anomaly field, the axis of which will frequently be at an angle to the ambient field, depending on the time of firing. In both cases, B_{earth} will be modulated by the anomaly field B to produce a resultant flux density, B_{tot}, that may have a magnitude greater or less than that of the earth's field. For archaeological anomalies, however, B is very much less than B_{earth}, typically 10 nT against 50,000 nT, so that this modulation will be minute. The value of B_{tot} actually measured then depends essentially on the type of magnetometer employed. If a single-sensor intensity instrument such as a proton magnetometer is used, it will measure the combined intensities (magnitudes) of the earth's field and B, i.e. B_{tot}, ranging between + and −10 in 50,000 of B_{earth} in the above example, as the dipole is traversed. We can now write, in terms of intensity,

$$B_{tot} = B_{earth} + F,$$
Eq. 3.4.

where the quantity F may be positive or negative with respect to B_{earth} and is known as the total-field anomaly. This is not the same as equation 3.1, which takes into account the directions of the vectors. The mathematical reason for this difference is that the sum of two vectors (as in eq. 3.1) is, in general, not the same as the sum of their individual magnitudes.

It is common practice to monitor the background flux density, B_{earth}, with the aid of local geomagnetic data or a second, remote, intensity sensor. Combination of the two sensor outputs constitutes a differential intensity (total field) magnetometer (fig. 2.2). The total-field anomaly is then found by subtracting the background intensity from that of the mobile sensor. Use of an intensity magnetometer sensor pair in gradiometer mode cancels out the effect of the earth's field almost completely. For a more detailed discussion of this topic, see Schmidt and Clark (2006).

In a vector magnetometer, such as a fluxgate or SQUID sensor, only a defined component of B_{earth} is measured, usually in the vertical z direction. Then

$$B_{ztot} = B_{zearth} + B_z.$$
Eq.3.5.

Thus a vector gradiometer, giving the difference between two such co-directional signals, completely eliminates the effect of the earth's field. For a single sensor, the anomaly components B_z and B_x, and the total-field anomaly F can be calculated depending on the angle of inclination, I, of the magnetic dipole axis from the horizontal. If the dipole is induced by the earth's magnetic field, this inclination will be identical with the earth's field angle of dip. In each calculation, the term $m/(x^2 + z^2)^{3/2}$ appears and is equal to m/r^3 as encountered above, confirming the rapid falloff of field strength with depth for the dipole. The three expressions are plotted in figure 3.4 for an angle of inclination of 70° north, for a dipole of magnetic moment 0.001 Am^2, buried at a depth of 1 m, and measured with a single magnetometer sensor that is carried 0.5 m above the ground surface (i.e., $z = 1.5$ m). Six main observations can be made about these anomalies.

- The response peak is shifted slightly to the south of the feature in the ground.

- A significant negative trough appears to the north of the feature.

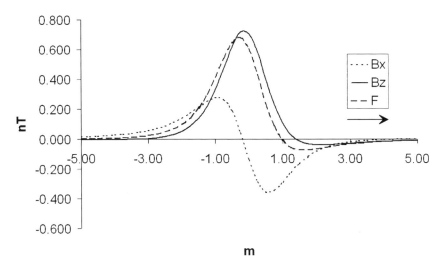

Figure 3.4. **Magnetic flux density components (B_x, B_z, and F) as a single sensor passes over a magnetic dipole at an inclination angle of 70°. The arrow indicates north.**

- To the south of the peak is a very shallow negative trough (becoming more pronounced for steep inclinations).

- The anomaly tends toward zero far away from the feature.

- The horizontal component B_x has a far stronger negative trough.

- Because the inclination in this example is fairly steep, the total-field anomaly F is very similar to the vertical component B_z.

This analysis confirms that plots of the data measured with a magnetometer cannot be taken as images of the physical reality of the subsurface. Instead, they are compounded by the geophysical nature of the readings, and this needs to be considered when interpreting any measurements. Data plots of the horizontal component represent the buried feature even less intuitively at this latitude. Furthermore, the angle of inclination at the point of measurement is a very significant factor in the anomaly shape, as can be seen through a simple explanation of the anomalies. Let us assume that the dipole at inclination 70° is traversed with a fluxgate gradiometer measuring only the vertical component, B_z, of the field (fig. 3.5). It can be seen, to the south of the feature, that the flux lines point in the same direction as the positive sensor axis, which will, therefore, measure a positive signal. To the north of the feature, this signal will be negative because of the opposing alignment. Where the flux lines are horizontal, and hence perpendicular to the sensor's direction, the instrument will record a zero measurement. One can also see that the southern part of the dipole is closer to the sensor than the northern, and, with the pronounced depth-dependency of the magnetic field, this makes the positive peak much stronger than the negative trough. Using the same principle, it is easy to explain the shape of the anomaly at other angles of inclination. At the magnetic equator, inclination $0°$, the dipole is horizontal, and hence the positive peak and negative trough have the same amplitude, the zero reading being directly above the dipole center. In the southern hemisphere, the N pole becomes shallower and the positive peak is, therefore, smaller than the negative trough. In all cases, however, the positive part of the anomaly is to the south, and the negative to the north. At the magnetic poles of the earth, inclination $+/- 90°$, the N and S poles of the dipoles are directly

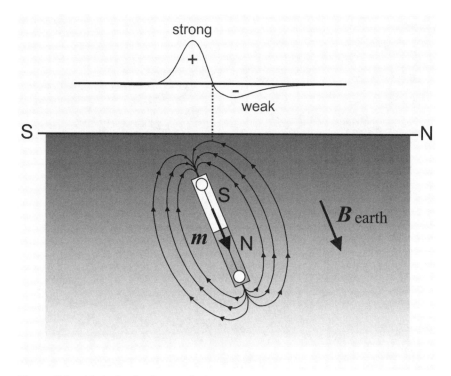

Figure 3.5. Variation in the vertical component B_z during the south-north passage of a fluxgate gradiometer over a buried magnetic dipole. The response form is characterized by the inclination angle of the induced magnetization.

above one another and will result in a single positive or negative peak, with smaller symmetrical troughs depending on the hemisphere.

Figures 3.6a and 3.6b show how the anomaly changes for different inclinations of the dipole, both for the vertical component B_z and for the total-field anomaly F. Unlike the vertical component, the total-field anomaly changes its polarity for negative inclination (e.g., for induced magnetization in the southern hemisphere), so that the positive peak appears there to the north of the feature. At equatorial latitudes, the total field anomaly is predominantly negative (Tite 1966).

The dipole response of the fluxgate gradiometer will essentially be smaller than that of a fluxgate single-sensor because the difference between two sensors, one above the other, is measured (see chapter 2). The effect of this configuration is shown in figure 3.7, where the responses of

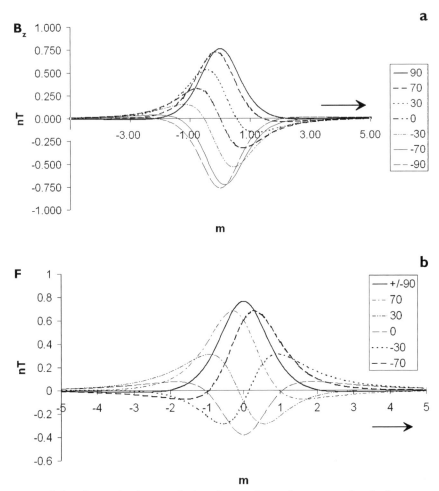

Figure 3.6. Anomaly changes during the south-north passage of a single-sensor over a magnetic dipole at different angles of inclination: (a) the vertical component B_z, and (b) the total field anomaly F.

different magnetometers are compared for the same dipole as discussed previously, i.e., $z = 1.5$ m. For the gradiometer, the lower sensor is at the same height as the single sensor, and the upper sensor is 0.5 m or 1.0 m above. The single-sensor response is obviously biggest, while the 0.5 m gradiometer peak is smallest, since the two sensors measure more similar anomalies than a 1.0 m instrument. Since the inclination is steep, the total-field gradiometer has a very similar response to the fluxgate gradiometer. The width

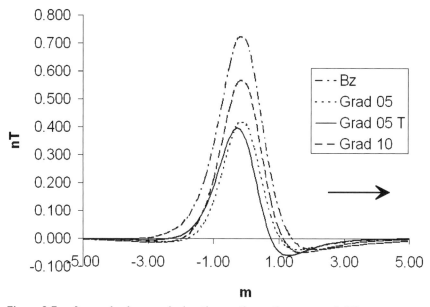

Figure 3.7. Anomaly changes during the south-north passage of different magne-tometer configurations over a magnetic dipole at 70° inclination angle; Bz is a sin-gle sensor measuring the vertical field, Grad05 is a 0.5 m separation fluxgate gradiometer, Grad05T a 0.5 m separation cesium-vapor instrument in gradiome-ter mode, and Grad10 a 1 m fluxgate gradiometer.

of the anomaly is best characterized by the peak width at half of its maxi-mum (full width at half maximum, or FWHM).

As can be seen from table 3.1, this width is considerably smaller for a gradiometer, and at this latitude virtually the same for the fluxgate and total-field instruments. As discussed in chapter 2, by taking the differ-ence between the two sensor readings, the gradiometer removes part of the background variations and hence sharpens the anomaly. Thus, for the same features, gradiometers will give better resolution but smaller

Table 3.1. Width and Center of Anomalies of Four Types of Magnetometers

	Fluxgate Single-Sensor	Fluxgate Gradiometer 1.0 m	Fluxgate Gradiometer 0.5 m	Total-Field Gradiometer 0.5 m
FWHM Width in m	1.52	1.37	1.29	1.27
Peak Shift to South in m	−0.20	−0.18	−0.17	−0.33

anomaly magnitudes than single-sensor instruments. Interestingly, the FWHM of the single-sensor anomaly is very similar to the depth of the feature below the sensor, and in fact this has been found to be generally true (Telford *et al.* 1990). The investigation also shows that for fluxgate sensors the location of the peak is closer to the real location of the feature in the ground at this latitude. A series of laboratory experiments carried out by Aitken and Alldred (1966) involved traverses of magnetic features beneath a stationary fluxgate magnetometer. These elegantly confirmed the above conclusions.

The above observations are made for a measurement traverse directly over the dipole in a north-south direction. What would the anomaly look like for other traverse directions? Generally, the observation point is now defined by $(x^2 + y^2 + z^2)^{1/2}$, where y is measured in the horizontal plane in a (magnetic) west-east direction. For a west-east traverse directly above the dipole, its two poles will be encountered at the same position. Consequently the recorded anomaly will be symmetrical. Taken together with previous results, this allows us to draw the vertical component of the magnetic flux density recorded over a buried dipole in two dimensions, i.e., the x-y plane (plate 2). The positive peak is accompanied by a negative halo, most prominent to the north, but which almost encircles the anomaly. At the magnetic pole, the anomaly peak and halo are entirely symmetrical. Away from the pole, this anomaly pattern is important when the best direction of traverse is considered for a geophysical area survey. In the case of a dipole, the most diagnostic traverse direction will be that of magnetic south to north, in that the characteristic signature of positive-zero-negative will be recorded (see above). The survey strategy, involving sampling interval and traverse spacing, could take this preferred orientation into account.

Anomalies of Arbitrarily Shaped Objects

In the course of magnetic surveying for archaeology, various features may be met that behave as magnetic dipoles. Iron fragments, which are often prevalent in agricultural and industrial topsoils, can behave as strong magnetic dipoles with a magnetization direction depending on the inclination of the earth's field and the shape and attitude of the object, and the presence, or otherwise, of remanent magnetization. Small fragments of fired clay, such as may be found on the site of a destroyed kiln or brick build-ing, will exhibit similar behavior. Small pits or postholes with a fill of or-ganic matter, cut, for example, in chalk beneath a significant depth of topsoil, may also be considered as magnetic dipoles. Other objects, with dimensions much greater than their burial depth are, however, more prevalent on archaeological sites. Linear features, such as walls, ditches, banks, and roads, and extended areas such as floors, ponds, storage pits, and building remains, will be encountered. It is, therefore, necessary to find a way of calculating anomalies for magnetic features of arbitrary shape. We are required to account for the fact that the measured flux den-sity at a point is the vectorial sum of the fields produced by all contribu-tory magnetic features in the ground. The dipole analysis is, clearly, a very specific and limited example of this procedure. We may, however, subdi-vide a buried magnetic feature into many small blocks, such as spheres or cubes, and consider each one as an individual magnetic dipole. Summing or integrating all the contributions, taking account of the directional as-pects of the components, leads to the resultant overall flux density for each measurement position.

A somewhat oversimplified example may help to illustrate the sum-mation procedure. A circular 1 m wide pit with a depth of 2 m is mod-eled as two spheres of 1 m diameter, one above the other, each with a volume-specific susceptibility contrast, with respect to its background, of $\Delta\kappa = 10 \times 10^{-5}$. Each sphere is represented by a dipole at its center. A magnetic sensor is placed 0.5 m above the top of the upper sphere. We can use the inverse-cube flux density approximation (eq. 3.3) to calculate the respective contributions of the upper (2.6 nT) and lower (0.3 nT) spheres. This yields an overall magnetic reading of 2.9 nT, clearly indi-cating the significance of burial depth to response. Interestingly, if the upper layer of the pit has a reduced susceptibility contrast from a more sterile fill, a very much reduced overall response is obtained, perhaps 0.5 nT, which may possibly be below detectable limits. The more practical example, of a homogeneous ditch of triangular section, has been evalu-ated by Heathcote (1983), and the results reproduced in figure 3.8. Here the relative contributions of the layers of the feature to the overall anom-aly are illustrated. It can be seen that 80% of the anomaly is attributable to the upper two-fifths (40%) of the section, and if the fill of the ditch is nonuniform, even more striking contrasts may be recorded. Heathcote

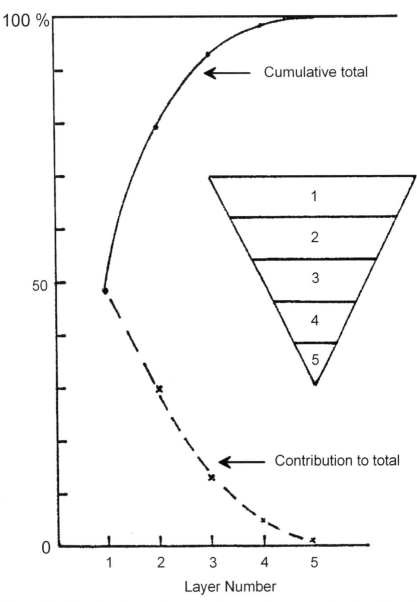

Figure 3.8. Contributions to the magnetic anomaly response in a 1 m fluxgate gradiometer from horizontal layers in a homogeneous triangular ditch of 2 m sides. After Heathcote, 1983.

also illustrates the decrease of response with burial depth to the top of a linear ditch of a 1 m triangular section. A line source of dipoles will give an approximate inverse-square falloff of response with depth of the field measured with a single sensor, a significantly weaker decrease than for a single dipole (i.e., point feature). Clearly the decrease is much more complex than this when measured over a great depth range using a gradiometer. The power coefficient of the decrease of flux density with depth is referred to as the structural index (Tsokas and Hansen 1995), and is regarded as indicative of the dimensionality of the feature (e.g., single dipole, extended ditch, Linington 1973).

A particular example of a feature with a specific structural index arises with a theoretical monopole anomaly. In this case, the length of the bar magnet modeling an isolated feature is very much greater than the distance to the point of observation. The magnetic effect of the more remote pole is then negligible, and the recorded anomaly is that of an isolated pole (S-seeking in the northern hemisphere) with a depth falloff, as seen in equation 1.2, inversely proportional to the square of the feature-sensor distance. Furthermore, the anomaly will show no reverse-field effect since the lines of magnetic flux pass radially to or from the isolated pole. It is a commonly held view that such monopole anomalies arise from isolated geological features, such as vertical igneous intrusions or dikes. In a less magnetic medium, they are unlikely to appear in an archaeological context. While an isolated weak anomaly may be erroneously identified as a monopole, it will be, in fact, a dipole for which the negative halo is below the limit of detection for the magnetometer used. On occasions, excavation of these monopole anomalies has confirmed the presence of shallow, isolated small hearths or small pits.

As well as discrete isolated features—such as pits and postholes and linear structures exemplified by ditches, walls, and banks—extended areas of activity are of archaeological interest. These include floor areas of demolished buildings, baked working areas associated with early industry, shallow middens, and old streambeds or paleochannels. Even isolated and linear features may have a significant lateral extent relative to their vertical dimensions, for example those of a wide Roman road. As mentioned before, such features can be represented by an assembly of individual dipoles, each generating its typical anomaly. The contributions from all dipoles are vectorially added at the location of the magnetometer, but with the strong

depth dependency of the dipole field $(1/r^3)$, the closer dipoles will contribute most. Hence an area of enhanced magnetic material creates higher
magnetometer readings, compared to the background, especially for thick
deposits. Although weaker, this effect will also be apparent in gradiometer
surveys, where the two sensors measure slightly different flux densities. Of
particular interest are the magnetic fields at the edge of the extended features, where the magnetic contrast with the surrounding material leads to
a strong peak. At the southern edge of the feature, the southern part of the
first dipole anomaly appears (as a slightly south-shifted peak). Over the
feature itself, all dipole anomalies combine to an elevated anomaly
strength, and at the northern edge the northern half of the last dipole
anomaly (its negative trough) is visible. As the feature's edges are essentially
two-dimensional, and the center predominantly three-dimensional, they
show different dependency with respect to the feature's thickness and
depth. Plate 3 shows the gradiometer anomaly of features with a 3 m × 3
m lateral extent, and a thickness increasing from 0.3 m (a thin sheet) to 3
m (a cube). The different influences of edge and bulk lead to clear peaks
over the sheet's edges (C), with a depression in the middle (fig. 3.9),
whereas the cube shows a fairly featureless overall anomaly (A). From this,
it is obvious that the shape of a magnetic anomaly can hold essential clues
about a feature's shape and size, but will convey less as depth increases.

Similarly, the anomaly shape can be used to estimate the depth of its
underlying feature. As noted at the start of this chapter, signal amplitude

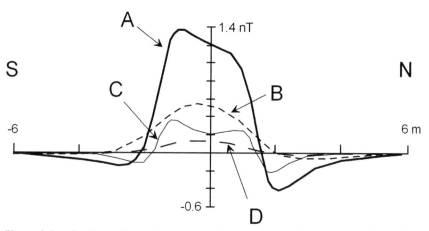

Figure 3.9. **South-north sections across the centers of the features of Plate 3.**

on its own is unsuitable for depth estimation because it also depends on the (normally unknown) strength of magnetization. However, the depth very strongly influences the width of the anomaly, and for a dipole the depth is approximately equivalent to the half-width (FWHM) of a single-sensor measurement (see table 3.1). It can generally be stated that a sharp and confined anomaly is related to a shallow feature, and a broad anomaly to a deep feature. Figure 3.10 shows how objects that produce the same signal strength at different depths and in different magnetization can still be differentiated by anomaly width.

So far, it has been assumed that the induced magnetization of a feature is always aligned with the direction of the earth's magnetic field. There are, however, circumstances in which a feature's shape can force its induced magnetization into a different orientation. This can be seen when magnetic surveys are carried out over buried iron pipes, such as are found with water supplies. Each section of a pipe shows alternate positive- and negative-end readings, regardless of the direction in which it lies with respect to the earth's field. This arises because, with elongated thin features of very high magnetic susceptibility, such as iron, the earth's magnetic flux lines are distorted so as to pass along the feature's axis and effectively create an induced long magnet with distinct N and S poles at its ends. A line of such pipes, jointed but distinctive, thus creates a beaded magnetic anomaly. An interesting study by Sowerbutts (1988) has demonstrated the

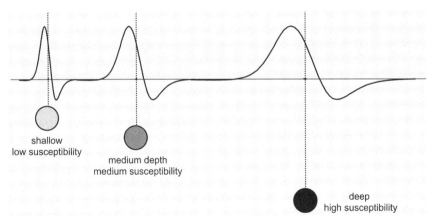

Figure 3.10. Effect of burial depth on anomaly width. Note that a deep feature of high magnetic susceptibility produces an anomaly of a magnitude similar to that of a weaker feature at shallow depth.

variability of line anomalies due to iron pipes of different composition lying in trenches at different orientations in the ground. This so-called demagnetization effect is due to the return field created by each individual dipole at the location of its immediate neighbors, which weakens the magnetic field that originally caused the induced magnetization. At the surface of such a feature, the inner dipoles have no counterbalance, and the shape of the interface hence determines the orientation of the overall magnetization. This effect is only noticeable for high magnetic susceptibilities ($\kappa > 0.1$, Telford et al. 1990) and especially pronounced for elongated features, both conditions being met by iron pipe and small but elongated pieces of iron debris. It should be noted that an igneous rock fragment detached from its parent will exhibit directional magnetization unrelated to the ambient field. All such anomalies can significantly complicate the interpretation of a magnetic survey, particularly if they cross an archaeologically promising area. They can usually be distinguished from archaeological anomalies, however, by the features described above (see igneous and ferrous responses, chapter 6).

Modeling of Magnetic Surveys

In measured magnetometer data, we may recognize a particular archaeological feature, for example a buried ditch, as a linear magnetic anomaly with a characteristic "fringe" of data of opposite polarity aligned with it. To obtain an archaeological interpretation, however, we would like to know the feature's shape and dimensions in vertical section, its depth and its composition relative to its surroundings. We therefore must model, theoretically, a structure such that magnetometer data calculated from it clearly resemble the appearance of the measured anomaly pattern. There are, obviously, a number of parameters to take into account in modeling archaeological situations, and some combinations of different values produce exactly the same overall anomaly shape. If we consider something relatively simple, such as a buried ditch, rarely will its fill be homogeneous, with a uniform susceptibility that contrasts with the surroundings. This is, perhaps, in sharp contrast with model studies associated with geological or engineering targets, where the magnetic material under investigation is often extreme in value, relatively homogeneous, and aligned along a known orientation. Furthermore, we have already seen that contributions to a magnetic anomaly decrease sharply with increasing depth in a way that is shape dependent. Early definitive studies by Scollar (1969) and Linington (1973), with the aid of then-novel computers, led the way to an appreciation of the variety of anomaly forms created by magnetic features of different shapes. They were followed by the work of Heathcote (1983), who produced an extensive collection of theoretical magnetic anomalies. These pioneering studies have been summarized by Scollar et al. (1990).

The problem of magnetic modeling in archaeological prospecting continued to receive a good deal of attention in the 1990s (Allum et al. 1995, Dittmar and Szymanski 1995). This decade began with Eder-Hinterleiter and Neubauer (2001) applying a specific magnetic modeling approach to a survey of a neolithic ditched enclosure. Their account contains a useful listing of methods employed by other current investigators, as well as a systematic analytical procedure based on the dipole-assemblage models of earlier workers. Their complete analysis entailed the collection of high-quality field data using a cesium gradiometer at a sample spacing of 0.25 m by 0.5 m, and the detailed examination of a limited number of excavated ditch sections. They took modeling one step further and automatically adjusted the model parameters in an iterative process to create the best possible fit between the calculated and measured magnetometer values, using typical magnetic susceptibility values from excavations at other sites as estimates. This process is usually referred to as inversion: using measured data to arrive at a model of subsurface features. A "few hours" of computing time were required to complete the analysis, and a picture very similar to an excavation photograph was produced. Although the real shape of buried features can only be confirmed by invasive investigations, their inverted geophysical results showed considerable potential in the archaeological analysis; they clearly indicated even the damage from continued agricultural plowing over the site.

Theoretical modeling and inversion of magnetic anomalies produced by archaeological features excited the interest of many other geophysicists as well (e.g., Tsokas and Hansen 1995; Dittrich and Koppelt 1997; Sheen 1997, Desvignes et al. 1999; Herwanger et al. 2000; Bescoby et al. 2004), and continue to do so. The necessity for this work arises from the sheer volume of magnetometer measurements now collected, and the possibility, in the absence of analysis by computer modeling, that the significance of their detail may be lost. While the practical complexity of often multiperiod structures continues to challenge complete identification, more advances should be a goal for future researchers.

In the study of likely or theoretical anomalies to be produced by buried features of archaeological interest, it is important to consider aspects of magnetometer data that will reduce certainty in the interpretation of data. Three important issues are resolution, noise, and sensitivity. There is interplay between these parameters, and they impact on all aspects of collection and interpretation of magnetometer data.

Resolution

As shown earlier, a very thin but laterally extended archaeological feature will be characterized by slight peaks around its edges, and this becomes significant when examining the resolution capability of a magnetometer. Resolution is defined here as the ability of an instrument to distinguish two closely spaced archaeological features as separate entities. For simple dipole anomalies, measured with a single sensor, the limit of resolution occurs when two equal anomalies are separated by a distance equal to their (common) peak widths at half-maximum height (FWHM). Clearly this definition becomes untenable if the anomalies are of different magnitudes or are buried at different depths, but the concept remains as a guideline. Another example of when it is important to separate two closely spaced anomalies is where ditches run parallel with each other. Sometimes the response from a large ditch can look remarkably similar to that of two closely spaced ditches (fig. 3.11). As we have seen, even in an idealized environment it may not be possible to achieve separation, owing to the underlying nature of the data. We shall encounter this problem again later, especially when analyzing field data. In chapter 2, we noted that a gradiometer system is of inherently higher resolution than a single-sensor device, because of its capability to act as spatial high-pass filter.

Noise

Apart from the physics associated with magnetic anomalies resulting from closely spaced features, two other factors must be considered when deciding upon which type of sensor to use for a specific application. A surveyor needs to understand what noise levels are likely to be encountered,

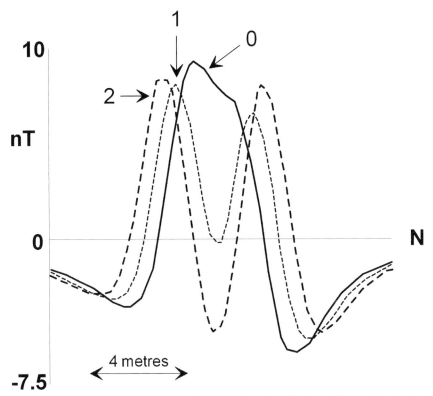

Figure 3.11. Magnetic anomalies of two triangular ditches of 2 m sides, buried 1.25 m below the lower sensor of a 1 m fluxgate gradiometer. The ditch edges are separated by 0, 1, and 2 m, and the inclination angle is 70°. Note that at 0 m separation the two ditches are readily confused with a single extended feature, but are clearly resolved at 1 m separation (after Heathcote 1983).

and what value of change in the earth's magnetic field a particular sensor type can detect.

By analogy with the phenomenon encountered in everyday life, a typical physical definition of noise is "any disturbance, especially of a random and persistent nature, that obscures or reduces the clarity or quality of a signal." The output produced by a magnetometer intrinsically carries noise from a variety of sources, together with the required signal, and the quality of the output is defined in terms of the instrument's signal-to-noise ratio. The required signal in archaeological terms is the magnetic anomaly arising specifically from a buried archaeological feature; all other signals together represent noise.

Magnetic noise can be divided into two major categories, incoherent or random noise, and coherent noise, and each of these categories can be further divided as follows:

Random Noise

1. *Instrument noise.* This is the inherent "white" noise produced within the electronics of an instrument. It is characterized by a broad continuous frequency spectrum that produces the typical hiss of an acoustic output. Although in early magnetometers the instrument noise was a very significant component of the overall noise level, in modern magnetometers (e.g., alkali-vapor) this has been reduced to an insignificant level relative to other noise sources. For fluxgate instruments, this still remains a noticeable limiting factor.

2. *Cultural noise.* Typically this arises from extraneous sources such as passing traffic, electric fences, and (some) radio transmissions. These sources of noise typically produce significant signals of variable amplitude and frequency, and are therefore often difficult to remove from otherwise coherent data sets. Their combined effect is to create a time-dependent background that may swamp the sought-for signals from buried archaeological remains.

3. *Operator noise.* Obviously, all operator-induced variations in magnetometer readings should be minimized, and conscientious field practice as well as thorough training is essential. Some operator noise may be consistent enough to be classed as coherent noise (see below), and it may be possible to reduce or remove it reasonably well. Others, however, will be fairly random and cannot be removed. Problems with instrument alignment, for example, can be more pronounced in strong winds or when surveying in high vegetation. Some operators have an uneven and inconsistent gait, and others may swing their arms while walking. All of these effects should be minimized through good field practice (see chapter 4).

Coherent Noise

4. *Instrument noise.* In addition to producing the random white noise described above, some magnetometers are prone to drift from an initial baseline with time. It is usually a problem resulting from varying temperature, mechanical shock, or vibration (Bartington and Chapman 2004). This form of instrument noise is easy to identify, and can be corrected if a systematic drift is assumed (see chapter 5).

5. *Diurnal noise.* In chapter 2, we discussed the presence of the magnetic wind from the sun, which creates periodic variations in the earth's magnetic field that are readily detectable with magnetometers. These diurnal effects are predictable only in their periodicity, and vary in magnitude and duration. They therefore represent coherent noise that can be eliminated from the magnetic record either by monitoring with a reference station or by surveying with a gradiometer, the sensors of which will pick up the diurnal variations equally with those of the earth's field (Weymouth and Lessard 1986).

6. *Soil noise.* The earth contains a range of magnetic minerals present to a greater or lesser extent near its surface; their variation can be significant, depending on past disturbances through environmental change or land use. These minerals are usually in the form of particles and fragments, and can be regarded as an assembly of tiny magnetic dipoles of varying composition and strength. Above an igneous subhorizon, they will have high magnetic susceptibilities and randomly orientated remanent magnetization. Otherwise they will exhibit induced magnetization, depending on their magnetic susceptibilities. Each member of this near-surface population will produce a small magnetic anomaly detectable by a magnetometer sensor. The total effect will be a confusion of magnetic signals—a level of magnetic noise, the magnitude of which will be highly dependent on the local surface geology. As a result, the soil noise over chalk upland will be significantly less than that over igneous terrain. Soil noise does not exhibit the time dependency of true random noise, and, on the microscale, is

repeatable and is therefore coherent. For the purposes of field survey, it can be regarded as the inherent background noise. However, variations in the ground of a nonarchaeological nature, such as refilled undulations from erosion and gullies, may produce more significant coherent anomalies that are readily confused with archaeological features. The leveled undulations of medieval ridge-and-furrow on many British rural sites are typical of such noise sources (Gaffney and Gater 2003). In volcanic soils, there will be considerable magnetic noise, and indeed the refill of ditches and gullies by a more "dilute" deposit will result in these features appearing as less-noisy anomalies. Typically, the range of soil noise varies from a few tenths of a nT, to several nT. More extreme examples of soil noise can be seen in zones of robber pits on archaeological sites (e.g., Herbich and Peeters 2006; Schmidt and Fazeli 2007).

7. *Cultural noise.* The presence of features creating unwanted magnetic anomalies that effectively are permanent are the most intransigent sources of noise in magnetic surveys. Thus power cables and pylons, stationary vehicles, wire fencing, iron-clad buildings and, if made of igneous rock or fired brick, standing walls are all detrimental to successful survey. Even when gradiometers are used, the presence of such objects is often detectable at a distance, making reliable survey difficult. More insidiously, buried ferrous and ceramic pipes and fired material such as brick fragments, used in site consolidation, may create intolerable levels of noise on an otherwise "clean" area. Occasionally however, the presence of iron "spikes" from the presence of small ferrous objects in the topsoil are diagnostic of the archaeological significance of a site. Evidently this is particularly true if the survey is of an industrial site, for example metalworking or other processes where high temperatures have been achieved. Small magnetic features also may have been introduced into the soil via manuring. In other cases, subtle but still significant ferrous patterning can be mapped. For instance, the site of a historic battle may contain ferrous relics of early military debris that may be identified

through the use of magnetometer surveys (see case study, chapter 6). Evidently establishing the significance of ferrous-type anomalies is not a cut-and-dried process; an archaeological interpretation still may be considered if the context of the noise is strong.

8. *Geological noise.* We have seen that a volcanic soil can produce a significant enhancement of soil noise. In particular, random elements of increased noise can be generated from naturally occurring magnetic cobbles. These can be difficult to distinguish from small pit or even ferrous-type responses, especially if the sampling density is low. However, more obvious geologically derived noise can be seen from coherent subsurface igneous material in the form of dikes and sills, which give rise to significant anomalies with forms dependent on the remanent magnetization orientation. The systematic or highly directional characteristics of such anomalies usually provide evidence of their nonarchaeological origin, as can their strength (fig. 3.12). However, the use of this material in early monuments, such as stone circles, must not be overlooked, and again the context of this form of noise must be critically assessed if the correct archaeological interpretation is to be made (see case study, chapter 6). Other coherent noise may be produced by the fill of palaeochannels in regions of early river flow. Although this variation in the past landscape can be mapped by many geophysical techniques, such as electrical imaging, some types of palaeochannels form magnetic responses that are very distinctive (Weston 2001). While some may see palaeochannels as unwanted responses in their data, others may find them a source of environmental data, as well as a means to identify locations of possible anthropogenic activity. Increasingly, this is a very important application area of magnetometer surveys in the light of interest in the mapping of early environments.

9. *Operational noise.* A significant enhancement of general noise can arise because of the presence of ferrous objects on and around the magnetometer operator. All operators should be

purged of such objects, if necessary by subjecting them to a pre-survey magnetic scan by walking around a stationary magnetometer (see chapter 4). When carrying out traverses of a site, it is also crucial that the instrument height above the ground remains constant. We have seen that, for a magnetic dipole, the instrument response falls off inversely as the cube of the distance, so that any variation of the sensor height will have a very significant effect on its response to surface features. This often produces a periodic coherent noise anomaly with a wavelength corresponding to the operator's stride. Also, if the instrument-response trigger is not activated at the same instant as the sensor is above a feature, there will be a consistent distance offset on a single traverse. If the operator then turns around, the return traverse will show an opposite offset, thus producing a shearing in the overall survey pattern, which can be likened to the blade of a wood saw. Finally, the act of starting and turning at the end of a traverse may give rise to errors in some magnetometers from their change of orientation. These will appear as anomalies at traverse terminations, and, so produced, will not be truly representative of site archaeology and must be classified as noise.

As we shall see later, the disruptive effects of some coherent, and to a far lesser degree random, noise may be reduced or eliminated by appropriate data processing. However, good field procedure, with the aim of minimizing noise problems in advance, is of crucial importance for the satisfactory outcome of a magnetic field survey. From the observations above, it is clear that instrument-derived noise is probably the least problematic aspect of noise associated with magnetometer data. It is, however, not possible to say which element of noise is the most significant—there are many factors that must be taken into consideration.

It is tempting to try to identify what noise level is, in fact, acceptable. However, the types of noise likely to be encountered during a magnetometer survey illustrate that levels can vary considerably and will depend upon geology, overburden, land use (both former and present), instrument, and experience of the operator. In short, noise must be largely judged on a site-by-site basis. Indeed, some elements of noise, such as an

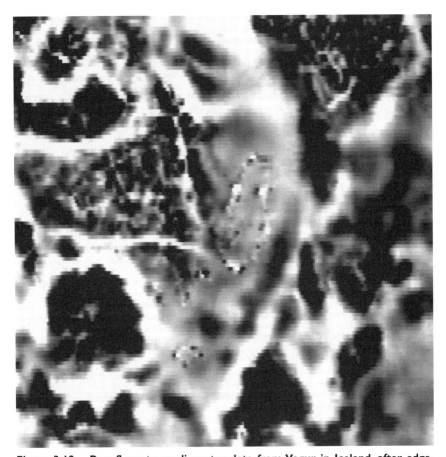

Figure 3.12.   Raw fluxgate gradiometer data from Vogur in Iceland, after edge-matching: 60 m × 60 m area; gray-scale plotted to white −100 nT and to black +100 nT, *north to top*. Igneous bedrock produced various intense anomalies ranging from broad responses up to 10 m across and in excess of ±300 nT, to areas of higher frequency anomalies less than 2 m across and in the ±50 nT range. These anomalies probably relate to internal structures and fracturing of lava. The most intense anomalies coincide with all but one of the mounds that surround the probable longhouse, indicating that the mounds are natural rises in the basalt (so-called lava tumuli). In addition, a weaker but distinct line of magnetic bipoles was detected over the outline of the structure visible on the surface (center of image). These were produced by igneous rocks that most likely had been used to face the turf walls of the building. Clusters of similar bipole anomalies were also detected a few meters beyond the eastern end of the longhouse, probably indicating the location of further archaeological features, such as pit-houses. Information and image courtesy of Tim Horsley.

igneous intrusion or a ferrous pipe, can be so strong that the summary statistics of the area may indicate high noise levels. But the noise must be
considered within the spatial extent of the unwanted anomaly. For example, in the immediate environs of a pipe or igneous intrusion, any magnetic response from a buried archaeological feature is likely to be
insignificant. However, away from such a major but spatially constrained
noise source, a similar feature may produce the same magnetic signature,
yet will be essentially mapped without problems; anomaly definition may
be good even when the statistics suggest otherwise.

In another situation, a weak response from an isolated archaeological
pit may not be distinguishable from the random noise generated from
topsoil containing dumped brick and ferrous material, even though the
noise level may be less than that identified in the previous example. However, a ditch producing the same strength of magnetic response as the pit
has a greater chance of being interpreted correctly from the overall shape
of the linear anomaly. This is also true of predominantly negative responses that are rarely of the same magnitude as positive responses on the
same site (see examples, chapter 6).

From these examples, it should be clear that noise levels, or signal-to-
noise ratios, cannot be the determining indication that an anomaly can be
identified or interpreted in an archaeologically meaningful manner. This
must be the case even if instruments capable of pT measurements are to
be used for magnetometer surveys. On a perfect site with very low background levels, the noise is likely to be several tenths of an nT.

Sensitivity

The smallest anomaly signal that an instrument can accurately recognize
and measure is often referred to as the sensitivity of the instrument. Thus
the true sensitivity of a magnetometer can only be measured using a calibrated magnetic source under laboratory conditions. Yet obviously the sen-

Table 3.2.   Sensor sensitivity levels. Note that values will differ between makes
of instruments and should be used only for general comparison.

Proton	Fluxgate	Overhauser	Alkali-Vapour	SQUID
0.1 nT	0.1 nT	0.05 nT	0.01 nT	0.00001 nT

sitivities of magnetometer sensors differ, depending on the science inherent in their measurement systems and the technology used to produce them.

Simplified Theoretical Model Conclusions

As we have just described, a major source of noise encountered in magnetometer surveys arises from the presence of unwanted magnetic features. These range from particles present in the soil surface to igneous geological outcrops, and are discussed in the following section. The effects of these noise sources are correspondingly diverse, but a view of this diversity and the role of the sensitivity of the instrument can be gained through the study of a simple model of the anomaly: a magnetic bipole simulating the range of features in terms of its change of length and depth of burial. In this way we can estimate what sensitivity means in relation to real burial parameters. Essentially we are interested in knowing how the measured signal varies with the target material (i.e., contrast), size of target, and depth (see fig. 3.10).

The model that we use to investigate the variation with depth derives directly from equation 1.3. The example is taken of a bipole magnetized along its axis by the earth's field at an inclination of 70° and traversed in a (magnetic) north-south direction. The model enables three significant parameters of the traverse to be evaluated, namely the positive peak magnitude, the peak width at half maximum (FWHM), and the separation between peak maximum and minimum as seen in the traverse. All these can be calculated as a function of the depth to the top (S pole) of the bipole below a fluxgate magnetometer recording the vertical component of the anomaly. Single-sensor and gradiometer modes, with 0.5 m and 1.0 m sensor separations and examples of 0.05 m and 1.0 m bipole length, are considered. Assuming a minimum distance of the single sensor to bipole of 0.25 m to represent an object at the soil surface, which gives 100% response, changes of the three response parameters are shown in figures 3.13 and 3.14, in which 0.05 m and 1.0 m bipoles respectively are illustrated. Figures 3.13a and 3.14a show the peak maximum on a logarithmic scale, figures 3.13b and 3.14b show the peak width at half maximum, and figures 3.13c and 3.14c show the maximum to minimum separation, all plotted as functions of depth down to 3 m. The two example lengths have been chosen to exhibit, in a simplistic way, the effect of feature size. Thus

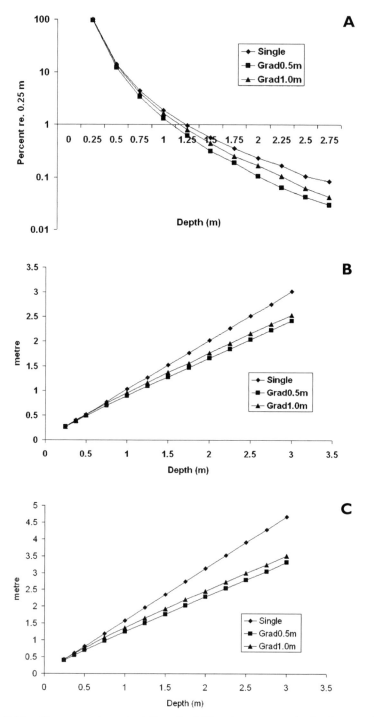

Figure 3.13. Response characteristics of a 0.05 m long magnetic dipole, at an inclination angle of 70°, as a function of depth beneath three sensor configurations: (A) peak amplitude, (B) peak width at half maximum (FWHM), and (C) separation of maximum and minimum amplitudes.

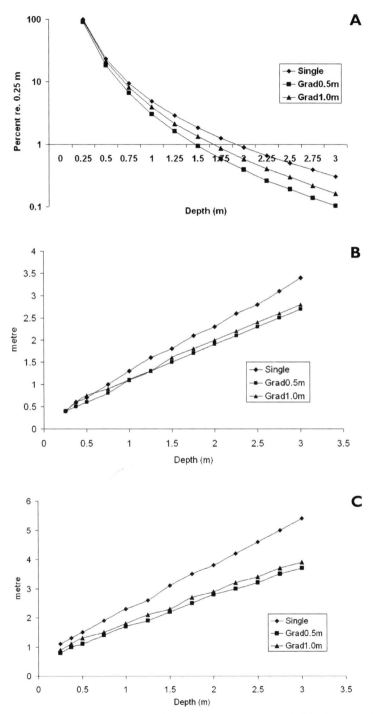

Figure 3.14. Response characteristics of a 1 m long magnetic bipole, at an inclination angle of 70°, as a function of depth beneath three sensor configurations: (A) peak amplitude, (B) peak width at half maximum (FWHM), and (C) separation of maximum and minimum amplitudes.

the 0.05 m bipole can be regarded as typically the "largest" small magnetic fragment in the soil, and is effectively a dipole at the depth distances chosen, while the 1.0 m bipole is reasonably representative of a near-surface igneous intrusion.

The observations that can be made from these plots are revealing. Taking the 0.05 m bipole as simulating a magnetized fragment, the detectable anomaly is effectively limited to the upper levels of the soil for all the fluxgate systems. A 10 nT response at the soil surface (0.25 m below the sensor) falls to 1 nT at a depth of 0.25 m below soil level, and, clearly, there will be a more rapid falloff for smaller fragments. Furthermore the half-peak width at the soil surface is 0.25 m, and, as originally pointed out by Aitken (1961), widens linearly with depth in a one-to-one ratio for a single sensor (see table 3.1). If we assume that two peaks of the same height can be resolved to their respective components if their separation is greater than their half-peak widths, it is easy to see how a uniform background can be generated by a large population of such fragments, with a superimposed "noisiness" when dispersed fragments of larger size are present. It is interesting to note that the depth falloff of the maximum-to-minimum peak separation is also linear for a single sensor, with a gradient of 1.5 (rather than the 1.0 found for the half-peak width). Either of these parameters may therefore be used in depth estimates for magnetic anomalies that can be regarded as dipoles.

As might be expected, the variation of response characteristics with depth shows significant differences with increasing bipole length, showing an increasing detectability, peak width and maximum-to-minimum distance. Interestingly the linearity of the last two parameters is maintained, but with increased gradient. In all cases, the presence of the second sensor of the gradiometer reduces the parameter magnitude. It should also be noted that the actual magnitude of the detected signal increases as the bipole is made longer, when its pole strength is maintained (i.e., the magnetic moment per unit length). Thus, at 0.25 m depth, the ratio of the 1 m to 0.05 m magnetic response is 3.2:1 for a single sensor.

Clearly, the response characteristics of the 1.0 m bipole can also be regarded as typical of those of "wanted" magnetic features. The peak half-widths, in particular, can give an insight into the resolution of two such features as separate. For example, two equal 1 m bipoles at a depth of 0.75 m below a single sensor may only be resolved if separated by a distance greater than 1 m. This has implications for the choice of the sampling interval for a magnetic survey. The variation in strength and nature of anomalies that can be measured by magnetometers is highly variable, depending on the amount of target present, how different the magnetic properties of the target are from the material in which it sits, and the distance it is from the sensor. The ability to define a magnetic response from a target also reflects the instrument that is used, and the sources of nonarchaeological noise that on occasion can swamp the signal. The ability to measure a particular signal is only the starting point, however, in understanding the archaeological interpretation of data from magnetic surveys.

DATA COLLECTION:
PRACTICAL PROBLEMS AND SOLUTIONS

The Changing Scale of Archaeo-Geophysics

As discussed in chapter 1, the first magnetic survey undertaken for archaeological purposes targeted areas where there was a high potential for locating buried kilns through their thermoremanent magnetization (Aitken 1958). In the years that followed, archaeologists broadened the use of the technique to investigate sites where contrast in magnetic susceptibility assisted in the location of features of interest (Heron and Gaffney 1987). These surveys were often small in scale and used to verify the context of a find or site. They were regarded by some as a cheap but inferior alternative to "real" archaeology (i.e., excavation). Over the past twenty years, however, the scale at which archaeological geophysicists have worked has changed out of all recognition. Their activities have advanced well beyond "wall-following" to site location and the mapping of whole cityscapes and landscapes. The most important single technique in this explosion of the use of geophysics in archaeological fieldwork during this time has been the magnetometer survey.

There are a number of histories on the subject that place the magnetometer into the overall context of archaeological geophysics (Clark 1990; Gaffney and Gater 2003). The common factor in these accounts is that the growth of magnetometry is fundamentally linked to instrumental and computational advances; evidently higher data acquisition rates require an increase in data processing ability. The fluxgate magnetometer in particular has a reputation for rapid data collection in a variety of field environments, and is regarded, when linked to computer-based data processing, as the "workhorse" of archaeological geophysics (Clark 1990). While there are common factors in how a survey is set out and geophysical data are collected, there can be very great differences in the density of data collection depending upon the scale of the investigation and the questions that are asked. From one perspective there is a need to find out as much as possible about what makes up a site, and from another there is a requirement to understand details of a site in its landscape context. Gaffney and Gater (1993) proposed three levels of investigation for archaeo-geophysical work that attempted to describe the ways in which projects are approached. The levels are:

Level 1—Prospection   Level 2—Evaluation   Level 3—Investigation

Some aspects of each level can be seen in table 4.1. The boundaries between these levels are subjective, but they allow some reference points in terms of strategies that have been developed through time in Britain. Here magnetometry has been embedded into the archaeological framework, and the above levels reflect the diverse use of the technique. It is possible to consider the roles of magnetometry in Britain as encompassing the location of unknown sites, mapping the context of previously or newly discovered locations of interest, and analytical use that attempts to avert the need to excavate. However, that is not to say that we can find out everything about a site via magnetometry; there are plenty of questions regarding dating, environment, artifacts, and fine detail relating to the sig-

Table 4.1.   Levels of Survey (after Gaffney and Gater 1993)

	Level 1 Prospection	Level 2. Evaluation	Level 3 Investigation
Technique	Scanning	Detailed survey	High-density detailed survey
Information	Areas of noise, archaeological potential, and individual strong anomalies, e.g., kilns	Delimiting and mapping archaeological sites and features	Detailed anomaly shape analysis
Archive	Field notes	Digital archive	Digital archive
Outcome	Possible Level 2 survey REPORT	Possible Level 3 survey REPORT	REPORT

nificance of a site that go well beyond the imaging of magnetic data. In fact, one of the greatest abuses of any remotely sensed data set is the over-interpretation of data. If one reads too many archaeological features into data, then it is likely that subsequent excavation by archaeologists will lead to the belief that the techniques are unreliable, which is of no long-term benefit to either surveyor or archaeologist.

Although the ultimate ideal is "100% detailed survey," in practice that is not achievable. Even those surveys that claim complete coverage of an area or site actually sample the area. Data are usually, but not always, collected in some type of grid format, where sampling points are measured along equally spaced lines (normally called traverses). Historically, the distances between sample points and the between traverses have been equal and, typically, 1 m. More recently, the traverse interval has become greater than the distance between sample points along the line. This is a compromise accepted many decades ago, although it is now usual to see much closer traverse intervals used for archaeological targets. The effect of sample density will be looked at later in this section, and it is important to know at this stage that data-rich surveys can be very expensive, due either to the need for additional sensors or to time considerations when undertaking highly detailed surveys.

Examples of the differing scales of investigation will be examined in chapter 6, but the concept behind them will be discussed now. As can be seen in table 4.1, the main difference in the Gaffney and Gater levels is the density of recorded values. In fact, in Level 1 (Prospection), it is possible to record no data whatsoever other than a visual impression, whereas in Level 3 (Investigation), the data may be collected along traverses separated by as little as 0.25 m or even 0.125 m. In effect, the use of geophysical techniques for archaeological purposes is a continuum where the variant is data intensity. It is probable that the majority of archaeological surveys can be regarded as Level 2, although the other two levels are increasingly common. It is usual to propose a strategy that cherry-picks from each of the levels, or a formal staged approach where the search becomes increasingly more intensive. In the supplement to the second edition of *Seeing Beneath the Soil*, Clark (1996) attempts to refine a Level 1 survey into a series of schemes that effectively cross the Level 1 and Level 2 boundaries; these two views are not really at odds with each other, they merely reflect different views of the same overall strategy.

In America, experience with archaeological prospecting has also led to a three-tier system for defining the scale of a survey, although the tiers are not equivalent to the three British levels. They have been described as detection, mapping, and integrity (Somers et al. 2003). However, in this classification system even the coarsest approach (detection) assumes that archaeological features are present and that detailed surveying is required throughout. In fact, there is a defined archaeological difference between detection and mapping in that historic or intensively occupied late-prehistoric sites fall into the first category, and early or nonintensive later prehistoric sites fall into the second. When a magnetometer survey is suggested, then there must be an implicit difference in the magnetic susceptibility contrast as a result of site use. The so-called integrity level would be used for the following purposes (ibid., 10):

1. To maximize the likelihood of detecting small, widely spaced, low-contrast features

2. To maximize the reliability with which the site is mapped

3. To assess the degree of preservation of the buried resource

Associated with Somers et al., 2003 is software available for downloading via the Internet at (www.denix.osd.mil/denix/public/library/ncr/culturalresources/mar05.html). This software guides users toward strategies that would be most suitable for projects within specific geographic locations (American Plains and mid-continent America). Evidently, this software is not suitable for all contexts, but it will lead new or inexperienced archaeo-geophysicists through the steps required to make a considered decision on the suitability of a strategy. Of course, experienced practitioners may disagree with some of the suggestions offered by the program, but software such as this attempts to make use of the common values and experiences of the user community in a productive way.

Planning Surveys

It is self-evident that as much information as possible should be gathered about the potential archaeological features at a site. For example:

- Is there excavation evidence that identifies potential features and the depth at which they are buried?

- Are there aerial photographs that show the type and coverage of potential features? Aerial photographs may also show problems associated with deep soils.

- Are there historic maps that relate to the area? It is important to understand the potential length of time that the site has been in use.

- How deeply buried are the targets likely to be? Is there a geotechnical or borehole report that relates to the pedological conditions at the site?

- What size of area are the remains expected to cover? Is it restricted by access arrangements, crop cover, or other limiting factors, such as modern incursions?

- Is there a known or supposed alignment to the suspected remains? If there is, then it may be appropriate to align data collection in a specific direction.

- What shape is the area within which the work is to take place? If it is too small or of awkward shape, then this may also affect the data collection strategy.

All of these questions can be addressed well in advance of a survey. Evidently some of them will have a major impact on the logistics of the survey. For example, a deeply buried site (more than 1 m overburden) might argue against the use of a standard 0.5 m separation fluxgate gradiometer. In such a case, a greater separation fluxgate system or a total-field instrument may be a better choice.

It should be feasible prior to a survey to find out if there are any service pipes or cables crossing the area. This information is likely to come from utility companies, real estate maps, or personal information from a landowner or farmer. All these sources have some value, but rarely will they be accurate. Experience has shown that not even the most diligent landowners know exactly where every pipe or drain has been buried, but they can usually give a good idea as to whether they are likely to be present. A large gas pipe or power cable can produce a distortion over a widespread area, perhaps 10 to 20 m either side of it, so this may influence the feasibility of a magnetic survey. This is particularly true when working within a narrow zone. Frequently an unexpected pipeline will produce the characteristic beaded response noted in the previous chapter.

If there is a road or a parking lot adjacent to the survey area and marked on a base map, it may be possible to have the area cleared of parked vehicles prior to the survey. A stationary car is likely to produce a magnetic anomaly that will be detected 10 to 20 m away, so removal of these sources of noise is very important. Passing vehicles will also produce unwanted and unrepeatable anomalies, so it is best to stop when one of significant size passes and check the magnetic disturbance it produces. To ensure high data quality, some traverses may have to be repeated.

Other considerations include the time of year the survey will take place. If there is some flexibility, one may choose to avoid any restrictive agricultural work that may be taking place. Tall vegetation can be highly frustrating when setting out a grid or collecting data. The magnetometer sensors can get caught, or there may be some difficulty seeing the direction of each traverse. It may be possible to schedule the survey after grass or a crop has been cut and removed; simply cutting it and leaving it on the surface can be problematic; laying out a grid can be very difficult in such circumstances. A good example of surveying under such conditions is illustrated in figure. 4.1. Images A and B are those obtained by fluxgate gradiometer surveys in successive years (2006–2007) over a prehistoric council-circle site at Sharp's Creek, in central Kansas, by David Maki and colleagues (pers. comm.). Excavation of such sites reveals that each consists of a central platform surrounded by four semisubterranean structures, and research suggests that they were ceremonial centers for their surrounding areas (Vehik 2002). The 2006 survey was carried out under adverse conditions; waist-high vegetation required that the gradiometer be carried at shoulder height. Data were collected, however, at a sample density of 0.125×0.5 m over an area of 60×60 m square to produce the image in figure 4.1a. In 2007, the site was cleared of vegetation and the area remeasured under normal conditions at a 0.125×0.25 m sample density, and extended to the east. The quality of the new image is striking (fig. 4.1b). It was estimated that the intensity of high-contrast features increased by 60% when compared with the 2006 survey, and many small or

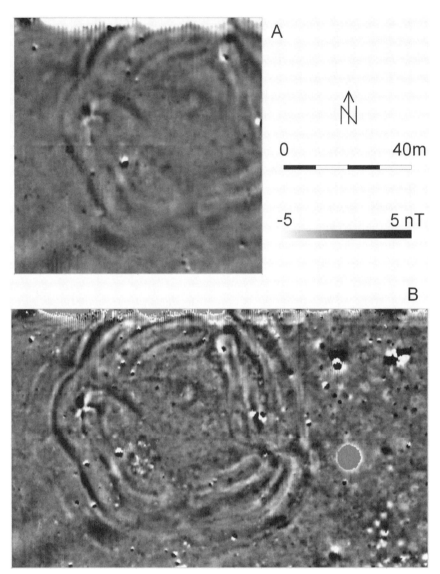

Figure 4.1. Images A and B are those obtained by fluxgate gradiometer surveys under different vegetation conditions at Sharp's Creek, central Kansas. The data in image B were collected after the vegetation had been cut down and illustrates the benefit of survey under optimum conditions. Images courtesy of David Maki.

low-contrast features were revealed. The implication is that the detail seen at the Sharp's Creek site suggests that council circles are much more complex than previously believed, and the value of good ground conditions is amply illustrated.

It must be obvious that bare earth is generally a good surface to work on, but, if the earth is plowed, then considerable additional noise may be added to the data. It may be best to leave the surface to weather, or wait until a crop has been sown. Similarly, if the surface is very muddy or slippery after heavy rain, then it may be difficult to collect data of good quality. If there is a need to work in all weather conditions, then good-quality, nonmagnetic waterproof clothing will be required, as well as sun hats and sunscreen.

There is little point going into the field if equipment is faulty or if the correct tools are not available. Surveyors need to ensure that they have sufficient nonmagnetic markers for a significant part of the grid. Even if a GPS system is used to locate the main points on a grid, 20 m or 30 m measuring tapes will still be required to divide the blocks into smaller areas. After repetitive use for setting out grids of a specific dimension, tapes are likely to wear in the same place, and once they have, they will stretch and break in a strong wind. As a consequence, it is good practice to take more tapes into the field than actually needed. The day before the survey, all batteries (magnetometer, GPS or total-station, and laptop) should be charged. Failure to do any of these can lead to major frustration, and even compromise a survey, if the time window is small.

Survey Preparations

Setting out Grids

One of the fundamental aspects of all geophysical surveys is that data are of no value if they cannot be accurately located in the field or on a map. There are many ways to ensure that a survey is "tied in" and properly documented. While it may be argued that the level of accuracy in relocation information may vary between surveys, depending upon the perceived outcome, the longer view of any survey requires that the position of any data point be recorded to the highest accuracy possible (Schmidt 2002). Essentially, there are two aspects to consider here. Firstly, if a grid is to be set up and data collected within that defined area, how should one go about establishing that grid with sufficient accuracy? Secondly, has this information been tied in to the surrounding landscape or on to a map with sufficient accuracy to allow the grid, and therefore individual anomalies located within it, to be relocated by another person, or for the data to be combined with other spatial data sets?

The site-coordinate system can be reliably and accurately positioned using a number of surveying tools. In most cases, the surveyor will use tapes, optical square, total-station (EDM) or GPS. Magnetometer data sets are usually contained within rectangular blocks, typically with sides of 10 m, or multiples thereof. In reality, this rigid grid will sit over an undulating surface resulting from ancient earthworks, modern interventions, and natural topography. It is therefore very important to document exactly how the grid has been established so that the magnetic anomalies within it can be relocated. Over short distances and relatively flat topography, it may be appropriate to position the grid using tapes and an optical square. If an optical square is not available, then right angles may be constructed with a tape using the 3-4-5 triangle rule. Alternatively, the surveyor may make use of the fact that the hypotenuse of a 10 m $\times$ 10 m block is 14.14 m, and in proportion for larger squares. Estimating a right angle this way is the least satisfactory way to position a grid, as tapes are notoriously difficult to use accurately in windy conditions or where the topography is rugged. An optical square is more accurate under these conditions, but distance still has to be measured using a tape—while the right angle may be reasonably accurate, the distance may not be satisfactory for survey. If tapes are used, they should be held as close to horizontal as possible, and checks should be made regularly to ensure that the grid remains within the required accuracy. This is particularly important on sloping ground where the slope and horizontal map distances can vary enormously (see fig. 4.2). An experienced surveyor will constantly check "along the line" to see if any points look out of place or if the line is diverging from the accepted alignment.

If a large area is to be surveyed, then tapes and optical square will not be adequate. In these cases, the grid should be set up either using a total-station or GPS device with real-time accuracy of a few centimeters. In these circumstances, it is acceptable to set out key points, perhaps 100 m intersections across an area, and tape in the positions to the corners of survey blocks. If common sense is used in checking the position of these corners with respect to one another, then acceptable accuracy will be maintained.

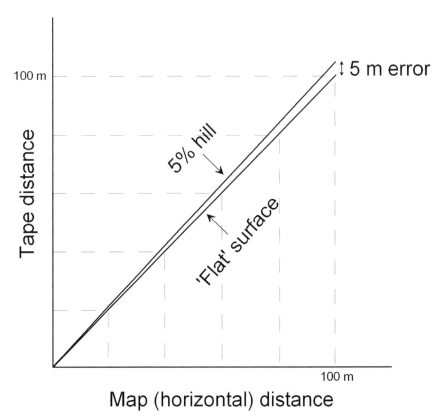

100 m

5% hill

'Flat' surface

‡ 5 m error

100 m

Tape distance

Map (horizontal) distance

Figure 4.2. A graph indicating possible errors when mapping and taping distances. These differences have implications for both data display and grid relocation.

If a total-station or real-time GPS is used to set out the grid, the advantage is that it will be tied in to the map with the same accuracy as all the grid corners, and that this tie in will be established at the same time. If a GPS is used, then no additional measurements are required. Should a total-station device be used, then a few measurements to features that exist on base maps and a few to any other obvious landmarks will suffice. If the magnetometer grid is adjacent to well-defined map features, then it is acceptable to use a tape to measure to these points. However, if the distance is beyond the length of an accurate tape measure, then a total-station or GPS system will have to be used.

Whichever method is used for setting out and tying in, it is imperative that the details be properly documented because attempting to recreate the grid with a different method can lead to significant errors. For example, if a grid is initially set out with tapes on a sloping site, then attempting to relocate it using a total-station will be unacceptable. This is true even if the surveyor has attempted to keep the tape horizontal during setting out. Grids that are set out, and therefore, tied in, by GPS should always be reestablished using the same technique. It is not acceptable to measure from features that are noted on a base-map to a grid positioned by GPS, unless the accuracy of the base-map has been established, since the location of features on base-maps can be wrong by several meters.

Of course, some magnetometers can be linked directly to a GPS device so that there is no requirement to establish a grid prior to surveying. At present these systems are in the minority, but their use will be described later.

Before Starting the Survey

Many modern sources of interference can be ascertained prior to any fieldwork via photographs or utility maps. However, some of these sources will not be apparent until the surveyor gets onto the site. Fences, both semipermanent and temporary, may not be included, even on recent maps. Clearly, these may be problematic in terms of the grid position relative to the sources of noise from brick or ferrous fence material. Where fences crisscross in corrals, paddocks, or livestock pens, data sets become too confused to make much sense; even temporary fences can produce an unwanted effect up to 5 m away. If fences cannot be removed, it may be best to use an alternative survey technique, such as earth-resistance, that does not rely on magnetic properties.

It may be worth spending some time clearing the area of any large ferrous objects on the surface. Depending on how urbanized a site location is, these may include metal sheets, pipes, agricultural items, and even wrecks of cars. If the data quality has to be at its highest, then sweeping the area with a metal detector prior to the magnetometer survey may be of value. Occasionally, modern material may have been buried within the area. This may be in the form of domestic or industrial objects; irrespective of its nature, this unwanted material will increase background noise, sometimes beyond that acceptable for magnetic surveying. In some instances, apparently perfect working conditions create a false impression;

magnetic material may have been used to landscape a piece of land prior to turf being laid.

Farm animals such as sheep, cattle, and horses should be moved away from a survey area; they have a tendency to pull up pegs and generally disrupt surveys. One aspect that should not be overlooked is the size of the area available for the survey. It is crucial that a surveyor understands that any magnetic anomalies are defined against the local background. If it is not known how the magnetic background varies, then it is impossible to judge what is significantly different, what is anomalous, and it certainly will not be possible to interpret the data in any meaningful manner. Wherever possible, a detailed survey undertaken in Level 2 and Level 3 investigations should be as large as possible to allow an accurate assessment of the magnetic background.

To collect data of highest quality, it is also important that the magnetometer operator be free of any ferrous metal that might otherwise influence the data. The checklist for this is long, but the most commonly missed items are eyelets in clothes and shoes, other metal parts of footwear (some rubber boots have steel stiffeners inserted into soles), nuts and bolts embedded in mud on the sole of a survey boot, magnetic strips on credit cards, zippers, belts, underwired bras, hairpins, paper clips in pockets, cigarette lighters, eyeglasses frames, and coins with a ferrous content. Even with all those items removed from the person who sets up an instrument, the magnetometer may be handed on for data collection to a colleague who is "magnetically contaminated." A quick test for ferrous metal on any operator is for one surveyor (known to be metal-free) to hold the instrument stationary while the potential operator walks around it as closely as possible without touching it. If the operator has any ferrous metal onboard, it will be obvious. It is always best to check this prior to data collection; repeating measurements is tedious.

Where a magnetometer consists of several parts with some ferrous content (e.g., battery packs, GPS), it is possible for the parts to produce a magnetic field picked up by the sensors. Although these components are mostly in a fixed position in relation to the sensors, their induced magnetization will partly be aligned with the earth's magnetic field. As a result, the effect on the sensors will depend slightly on the orientation of the whole assembly relative to magnetic north. It is therefore important to es-

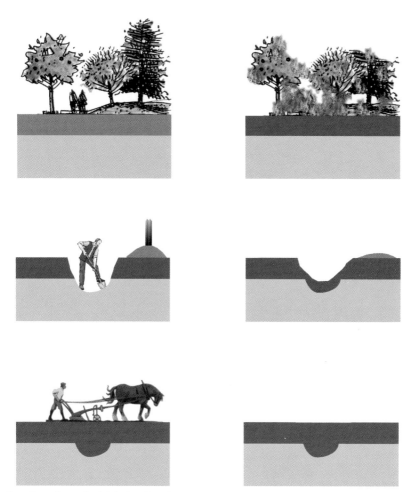

Plate 1. Creation of a high (positive) magnetic anomaly by excavation and gradual refilling of a ditch. Magnetic susceptibility of the topsoil is greater than that of the subsoil from anthropogenic and pedogenic processes (settlement and fire, top diagrams).

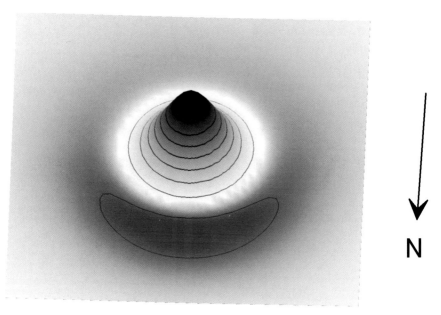

N

Plate 2. Two-dimensional (x, y) anomaly characteristics above a magnetic dipole at 70° inclination angle. Note the symmetry east-west. Here the positive anomaly is black.

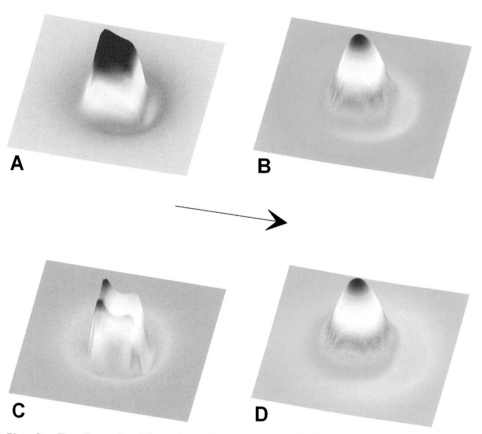

Plate 3.   Two-dimensional fluxgate gradiometer anomalies from a magnetized square feature of 3 m sides: (A) 3 m thick and 0 m deep, (B) 3 m thick and 1 m deep, (C) 0.3 m thick and 0 m deep, and (D) 0.3 m thick and 1 m deep. In each case the magnetic sensor was 0.3 m above ground. Note the vertical scale differs for each image.

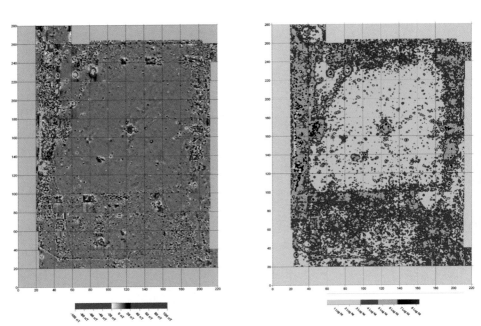

Plate 4.   Standard Deviation Analysis. Figure, *north to top*, shows data from a nineteenth-century U.S. military site. *Left*, minimally processed data. *Right*, converted to a variance map results from an analysis window with 1 m radius show neighboring bipoles (from brick, iron, or steel) joined for clearer outlines of active areas. Rectangular shapes particularly visible to the south are probably building rubble. Images courtesy of Lew Somers.

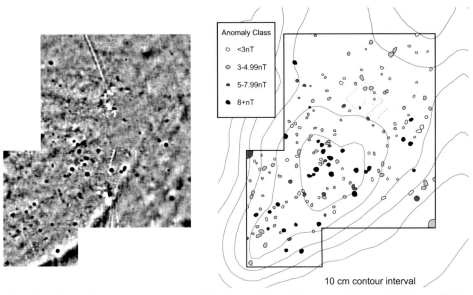

Plate 5. Magnetic anomalies at a prehistoric native settlement at Brown's Bottom, Ohio. *Left,* part of the site, in an image gray-scaled from −1.5 nT white to +1.5 nT black. *Right,* a plot of discrete anomalies, superimposed on the topography, identified as pits and ovens. Images courtesy of Jarrod Burks.

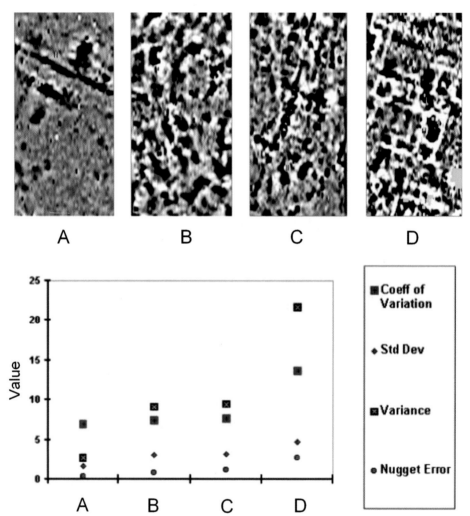

Plate 6. Statistical analysis of data from four areas of Roman-era Wroxeter: (A) background, (B) industrial, (C) artisan, and (D) elite. In each, the data set was stripped of magnetometer anomalies, and the low-level signal analyzed using various measures of variability. It suggests coherent statistical data may be held within the background that relates to uses of zones, and that templates can be built to help interpret use within the rest of the data set from the site.

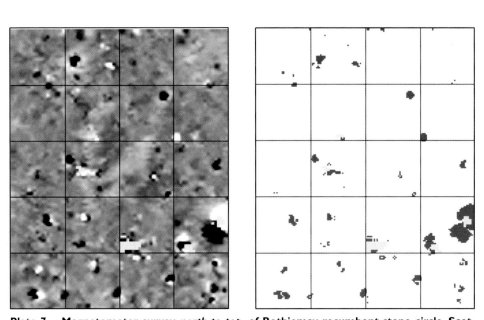

Plate 7. Magnetometer survey, *north to top*, of Rothiemay recumbent stone circle, Scotland. *Left*, an image of interpolated data (gray-scale −10 nT white to +10 nT black). Note dummy readings at sites of standing stones. *Right*, an elective palette presentation of data (+10 to 9.5 nT red, and −9.5 to −10 nT blue) to isolate strong igneous anomalies.

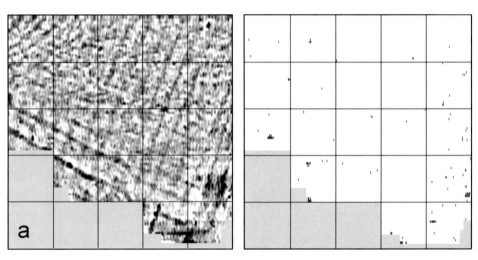

Plate 8.   Magnetometer survey, *north to top*, at Towton battlefield. *Left*, an image of interpolated data (gray-scale −0.5 nT white to +2 nT black). *Right*, an elective palette presentation (+1.3 to 1.2 nT red, and −1.2 to −1.3 nT blue) to show discrete ferrous anomalies (spikes).

tablish, prior to survey, how far away from the sensors these parts have to be to minimize their unwanted effect.

Setting up Instruments

At any level of investigation undertaken (see above), it is fundamental that the equipment be maintained, set up, and used appropriately if the often-minor magnetic anomalies of archaeological features, described in chapter 3, are to be measured and mapped. Magnetometers that measure the total-field require little in the way of setup, apart from following the manufacturer's guidance on assembling the system and selecting sensor orientations that minimize heading errors. However, all instruments benefit from warming up to a stable operating temperature before use, and it is good practice to attach the sensors to the electronics and switch on the system twenty minutes before any setup or measurement-recording is attempted.

Fluxgate Gradiometers in General

The most common instruments used for archaeological magnetometer surveys are fluxgate-based, and most of these require careful setup; this ritual must be observed if data of sufficient quality are to be collected for an archaeological investigation. There are currently three commercially available fluxgate gradiometers that are used worldwide for archaeological investigations; of these the 0.5 m-separation Geoscan FM series and the 1.0 m-separation Bartington Grad instruments are commonly used, with Foerster Ferex instruments—which employ sensors of varied lengths—gaining popularity. These instruments essentially use the same sensor technology, but their approaches are very different. While the popular Geoscan instruments require a manual setup, the Bartington system provides an automated procedure, and the Foerster gradiometers are calibrated in the factory and require no further setup in the field.

The objective of setup procedures for the Geoscan and the Bartington instruments is to align the sensors so that they give the same reading when rotated around their vertical axis. Slight rotations are hard to avoid when walking along transect lines, and the impact on data acquisition has to be minimized. Obviously, when turning the instrument to walk lines in opposite directions, the rotational effect is largest, which is why parallel traverses are often preferred, especially for data collection by inexperienced operators.

First, a location has to be selected where the setup of instruments can be undertaken. It is best to find an area relatively free of magnetic anomalies. Once the instrument has warmed up, a zone of low magnetic background must be found by scanning an area while keeping the head of the instrument in the same orientation. At this point it is imperative that the instrument be kept vertical or the display will register variations from misalignment of the sensors, rather than deviations from buried features or the soil. When a zone is found where there is little change in response, no matter what the value is on the instrument's display, it is likely to be a suitable spot to set up the instrument. The variation should be no more than about 1 nT, and the area around it should extend for 2 to 3 m. At this stage, it will not really be possible to find areas with less variation, because the instrument is not yet set up.

There is no reason why this position has to be in the survey area, but it makes sense for the setup to be central, because it will become a reference point regularly checked during the day. If at this stage the sensors are very badly aligned, it can be worth setting up in a least-noisy area, then scanning around to find a better, or more convenient, position and going through the setup a second time. Whatever point chosen for reference should be marked with a plastic peg or wooden stake. Some surveyors mark magnetic north on the ground at this stage—with paint, pegs, or canes—to ensure a standardized setup later. If a "quiet" point cannot be found, there are still a number of possibilities. Firstly, the surveyor might go off-site to set up. In one instance, while surveying in a disused railway freight yard, where the target was buried rolling stock, a reference point for setup was established in a nearby park. Secondly, some practitioners have suggested that a plastic box or nonmagnetic stool be used to raise an instrument away from surface noise. If an instrument can be raised to about 2 m from the surface, then a relatively uniform magnetic-field reading will be reached (www.cast.uark.edu/nadag/educationalmaterials/mag/tuning-fm36.htm). This technique may be useful where the soil is of igneous derivation or exhibits high magnetic susceptibility changes for other reasons.

During the whole setup procedure and during data collection, the instrument must always be held vertically, otherwise the sensors will measure a different, undefined component of the earth's magnetic field.

Geoscan Fluxgate Gradiometers

In the case of Geoscan FM instruments, the following procedure should be undertaken after a 15- to 20-minute warm-up. It consists of two parts: (A) the balancing of the electronics for the two sensors, and (B) the physical alignment of the two sensors.

To achieve accurate compensation of the earth's magnetic field in the gradiometer, ($B_{lower} - B_{upper}$, see chapter 2) the electronics have to be set up identically for both sensors in the instrument. The best way to test this is to turn the instrument upside down: if the electronics are well-balanced, the reading in this orientation will be the same as in the upright position; if not, the balance can be adjusted with a nonmagnetic tool. For an accurate comparison of these two orientations, the instrument should be rotated in such a way that the two sensors exactly interchange their positions. The rotation-and-balance adjustment is to be repeated until the difference between the two readings is no more than 1 nT. Obviously, viewing the instrument's display when it is held upside down is difficult, and users have developed their own methods to overcome this problem. One is to hold the instrument overhead when inverted, although this means that the sensors are then above where they will be in normal orientation. Another is to press the zero-button of the instrument while it is inverted at the usual ground height. Although the display is temporarily unseen, the reading is reset to zero. However, this method is not recommended by Geoscan, which has updated the design of its FM256 model, giving it a button that allows the operator to "hold" the instrument display of readings from inverted orientations.

To make a gradiometer's reading independent of even small rotations about its vertical axis, it is important that alignment of each of the two fluxgate sensors comprising the gradiometer be as accurately vertical as possible. Ideally, this requires that each sensor be aligned with two axes, so that in total four axes would need to be manipulated for the whole instrument. It is generally accepted, however, that adjusting the sensors such that they are parallel to each other and "reasonably vertical" is sufficient. This only requires one adjustment axis for each sensor if the adjustment axes of the two sensors are perpendicular to each other. To adjust the sensors therefore, the following procedure should be followed. While pointing the gradiometer north, the operator memorizes the instrument reading and then turns the instrument south for a second reading. If the sensors are vertical, the two readings will be the same. Any difference between the two is halved by using a north-south sensor-adjustment screw. North and south readings are then checked again, and the adjustment is repeated until the two values are within 1 nT. The same is then repeated for the east and west, adjusting the east-west screw of the other sensor. Unfortunately, adjusting one direction also influences the alignment in the other direction because the sensors are "only" parallel (see above), and so adjustments have to be repeated several times in turn, until north-south and east-west readings are each within 1 nT and, ideally, all four directions show the same reading.

With sensors now well-aligned, the initial electronic balancing (see above) should be repeated, because any slight rotation of the previously unaligned sensors, when inverted, may have introduced errors. As a final check, the four directional readings should be tested again in normal orientation and adjusted if necessary.

The setup location will also act as a "zero point" when the "log zero drift" option of the instrument is activated. In order to use this function, the instrument display is set to zero after balancing the electronics and aligning the sensors, and before commencing the survey of a survey block (or data grid). Once a grid has been completed, the operator returns to the zero point to record the instrument's displayed value (i.e., to log the drift from the initial zero reading). This additional datum value enables the software to correct for instrument drift when later downloading the data.

Although the setup procedure appears laborious, it ensures that data of the highest quality can be collected. This has to be the primary concern of any surveyor since postacquisition data-processing can never correct for all the small inconsistencies introduced through poor instrument setup. Although Geoscan instruments maintain their stability relatively well during a day of surveying, regular checks of the instruments at the setup point, with small adjustments if necessary, are highly advisable.

Bartington Gradiometers

For Bartington 1.0 m instruments, there is an automated procedure that adjusts the gain, offset, and alignment of the gradiometer very effectively and simply (Bartington and Chapman 2004). The operator is guided through a rotation sequence, during which each sensor is adjusted automatically for minimum error. Assuming that a reasonable setup position has been located, rotational errors may be less than 0.5 nT. The setup takes less than 10 minutes and requires little input from the operator. One of the instrument options is a pair of fluxgate gradiometers separated laterally by 1 m to facilitate speedier data collection, and extra care must then be taken to locate a setup area large enough to accommodate the full instrument. To facilitate optimum alignment, a nonmagnetic tripod is available. The gradiometer locks into the tripod for stability during alignment, and the tripod can be extended to vary the distance of the sensors from the ground.

Collecting Data

Survey Direction

With a grid set out, in which direction should the data collection be undertaken? There are three schools of thought as to the orientation of a traverse. Firstly, since most model data available are plotted as north-south traverses, there is considerable merit sampling the north-south response of a magnetic anomaly with as many data points as possible, because the density collected along the traverse line is greater than that between traverses. It may be, though, that some small anomalies will be missed if the spacing between traverses is greater than the sampling interval. Secondly, there is merit in walking east-west because data presented in Geoscan's Geoplot software orients the grid north to the top of the data plot, in common with cartographic convention. While this may be of little relevance if data are to be shown in plan form only, there are significant advantages if an X-Y plot is produced. For fluxgate gradiometers, both options have the advantage of an instrument operated in one of the directions in which it was set up (see above). Or, thirdly, fit the magnetometer grid to the survey area and walk the traverses in whatever direction is easiest, while maintaining a constant direction with the magnetometer to reduce the influence of any heading error. This last option is the easiest, and most surveyors will follow it, although they will normally prefer collecting data either toward grid north or at right angles to it. In reality, there is no "correct" orientation, and, as the sample density of data collection becomes greater, theoretical argument on the issue fades in importance.

Sampling Strategies

Gridding

Traditionally, survey data are collected by means of a defined raster-like grid, subdivided by traverses and measurement points, sometimes called stations, taken along these lines. The main advantage of this approach is consistent and unbiased sampling of a whole area. No matter the size and shape of an anomaly, it will always be sampled in exactly the same way throughout the survey grid, and interpretation of anomalies will be consistent. The sampling may be too coarse or too dense, but it will always be the same. In order to achieve this, the whole area has to be subdivided into manageable blocks, or data grids, prior to the survey, which can be a time-consuming task. Data collected in this way can easily be displayed with each data point represented by a rectangle of a size representing the raster-grid resolution.

Randomizing

An alternative approach is to record magnetic data simultaneously with the location of each sampling point. Although not commonly done at present, a high-precision, real time GPS device can be attached to a magnetic measurement system to record the location of each data point automatically. The system can either be carried or pushed by the surveyor or towed behind a suitable vehicle. The great advantage is that the survey area does not have to be set out prior to measurement, and position data are acquired at the same time as magnetic data, considerably reducing time and effort. The data point positions are considered to be random (i.e., not on a regular grid), even if the surveyor attempts to collect them fairly systematically.

There are various issues that need to be considered when adopting this approach:

Data coverage. Since data collection need no longer be governed by defined survey lines, and even experienced surveyors can find it difficult to walk along equally spaced and straight lines, let alone drive them on surveys, the GPS random approach beckons. Fortunately, as well as tracking where a measurement has been taken, modern GPS systems can provide guidance displays to lead surveyors along defined paths as well. These displays come in either the form of automatic direction-indicators that help walking or driving in straight lines, or in full displays showing where gaps still exist in survey areas already covered. These approaches have already been adopted in precision agriculture, and some of its GPS systems can be directly adopted for archaeological geophysical surveys. A vehicle-mounted fluxgate system by Foerster (Multi Cat) marks the ground with white foam to indicate the areas it has surveyed, providing completion-guidance for its operator in this way.

Data presentation. No matter how carefully data collection is guided, the resulting distribution of measurement positions will not represent a regular grid; it will be random. When data are collected at high sampling densities, e.g., from surveys with high random sampling, the data may be re-gridded into a regular raster pattern. The calculated raster array normally does not reflect actual measurements (i.e., value and accurate position) but uses instead estimated data values at chosen intervals. Since survey lines are no longer exactly straight, or even parallel to a raster grid orientation, it is common to interpolate the randomly collected data to a raster with square cells of a size somewhere between the approximated traverse spacing and the sampling interval. Careful consideration has to be given to areas where random sampling has been too sparse to be interpolated, and that should be left empty in any generated raster pattern. A completely different approach is to maintain the random position of the samples, and represent the areas between measured points through a triangulated network of polygons. Sauerländer et al. (1999) suggested using Delaunay triangulations and their associated Voronoi diagrams for mapping, which correspond to the use of nearest-neighbor interpolation on a fine grid (e.g. 0.05 m × 0.05 m). The resulting polygons will honor the original sampling regime. If smooth transitions between data values are required (i.e., interpolation), natural-neighbor gridding can be used to remove data mismatch at polygon boundaries (Li & Götze 1999).

Sensor arrays. If multiple sensors are combined into an array and a GPS tool is used to monitor the position of data acquisition, it is important to calculate the exact location of each sensor from the GPS readings. For example, it is necessary to take into account the direction in which the array is being moved, and then determine where the left and right sensors are. The best way to monitor this is to calculate a velocity vector from sequential GPS positions, and, for further positional accuracy, pitch and roll of the array can be recorded with separate sensors to make necessary adjustments.

Scanning

A variation on random sampling is scanning, which refers to non-recorded assessment of a site with magnetometers. For example, an experienced surveyor can walk along widely spaced straight lines observing the instrument display and putting markers in the ground whenever a significant reading is encountered (Gaffney & Gater, 2003). The result of this sort of investigation is a set of ground markers that may be used to assess the archaeological potential of a site. The limitations of such evaluations have to be weighed carefully against the greatly reduced effort of this Level 1 strategy. It will be impossible to interpret individual anomalies, distinguish between modern and archaeological features, or link features together. Scanning also depends on the judgment of the surveyor to decide which readings should be marked, and is hence prone to the subjective biases and preconceptions of a site held by an operator. Good practice requires detailed Level 2 and Level 3 surveys to establish the origins of the scanned anomalies.

Sampling Intervals

Although authors sometimes describe their magnetic surveys as "complete" or "detailed," there is always some form of sampling taking place during data collection. Analog recording systems, such as those described in Clark and Haddon-Reece (1973), are continuous along a traverse, yet there are still gaps in the data collection between the traverses. Other systems rely on sampling at a defined instrument rate, usually expressed as samples per second (Hertz), and the resulting sample density then depends on the speed of instrument movement. In gridded surveys, this instrument speed must be kept constant to ensure that the average sample density (calculated as the length of a traverse line divided by the number of points along the traverse) is an accurate representation of the actual sample density. This requires skilled operators, and can be difficult in awkward terrain. In random-sampling surveys (i.e., with GPS-recorded positions), this is far less of an issue, and corrections can be applied, although interpolation errors may still be noticeable. Overall, however, the maximum in-line sampling interval only depends on instrument speed, and seldom limits the overall spatial survey resolution.

More difficult is the choice of spacing between adjacent survey lines, which can only be improved through more traverses or more sensors in an array, both options being expensive. Line-spacing therefore normally governs the perceived spatial resolution of a survey. It was, for example, shown on an Iron Age test site that a 0.5 m × 0.5 m resolution survey allows far better identification of archaeological features than a 0.25 m × 1.0 m survey (Schmidt and Marshall 1997). The latter should not, therefore, be referred to as high-resolution survey, even though the sample density along the survey lines is relatively high.

What, then, is the right choice of survey resolution? There is no single answer to this important question since there are many aspects that should be considered. In any case, it will depend on the archaeological questions to be answered by the investigation, which can partly be defined by the survey levels discussed earlier. The following three areas of interest may be important in the choice of sample density.

Feature Sizes

If the aim of the survey is to identify the presence or absence of archaeological features (Levels 1 and 2) it may be sufficient to obtain a single reading across an archaeological anomaly. The coarsest sampling resolution (i.e., usually the traverse spacing) would then be roughly the same as the width of a typical feature. However, if a detailed analysis of the magnetic anomaly is needed (Level 3), for example to evaluate the depth and shape of the archaeological remains, an adequate number of data points are required for each anomaly. If data are degraded through noise (e.g., instrument noise), a higher sampling density can also help to recover the shape of anomalies. This idea of matching the measurement resolution to the stated survey aim is illustrated by a survey at Teotihuacán, Mexico, where a magnetometer survey with 2 m x 2 m resolution, normally deemed too coarse for archaeological features, was used to detect a large underground tunnel (Chávez et al. 2001).

In a so-called "staged survey strategy," magnetometer surveys gradually progress from lower-density collection to higher (and equate with a move up in levels), based on the analysis of interim survey results. For example, having identified vague indications of archaeological features in a coarsely sampled data set, detailed high-resolution surveys can then be targeted to small areas over these features and efficiently deliver all necessary information (Vernon et al. 1998). It has to be remembered, though, that coarse surveys may miss small anomalies altogether, and a detailed survey from the outset could be a more desirable option. That is not to say, if data were reinvestigated after initial survey aims were fulfilled, that measurements recorded at higher sampling densities from the beginning would lead to new insights not previously considered.

Feature Depths

It was shown earlier that, with increasing depth, the width of an archaeological feature's magnetic anomaly increases while its strength decreases. It is therefore important to consider the implication of a feature's depth on the necessary sampling resolution: shallow features need a higher density. The extreme cases that are commonly encountered are very broad geological anomalies and very narrow spikes from ferrous fragments in plowed soil. Prior information about suspected buried materials is very useful when determining survey resolutions.

Practitioners have developed a rule of thumb for the selection of an appropriate spatial resolution for a survey: it should neither exceed the size nor depth of the archaeological features. For surveys of typical remains and normal topsoil depth, this would suggest a sampling interval of 0.5 m × 0.5 m. Given the recent developments in magnetometer systems, this has become an achievable target.

Over-Sampling

Given the discussions above, it may seem odd that some practitioners wish to over-sample; that is, collect more data than is required to define an anomaly. However, in some soil-science applications, very small changes in magnetic soil composition may be of interest in themselves (Mathe and Leveque 2003); in archaeology they are normally referred to as soil noise (see chapter 3). If a high-sensitivity magnetometer survey is undertaken with high spatial resolution (e.g. 0.2 m × 0.2 m), it is possible that this soil noise will be apparent in the data. There are two schools of thought with regard to this:

1. If soil noise shows in the data as individual anomalies, it is easier to distinguish from instrument noise, and therefore

anomalies of archaeological origin can be identified more clearly as such.

2. Too many small-scale variations in the data may distract from major archaeological trends and make an overall assessment (the bigger picture) more difficult to obtain, hence over-sampling should be avoided.

Both of these arguments have their merits, and it depends on the strength and extent of soil noise, as well as on the archaeological anomalies, as to which of these two arguments is the more relevant.

CHAPTER FIVE
PROCESSING AND DISPLAYING DATA

In the preceding chapter, the focus was on designing data collection strategies that, when accompanied by a robust approach to field procedures, should lead to good-quality magnetometer data. However, even the most accomplished field operator cannot produce data completely free from defects that result from either the data acquisition strategy or the particular instrument used. In this chapter, suitable approaches to computer processing will be described; the algorithms discussed will not restore inherently flawed data, but should produce images of data that will be possible to interpret in a meaningful way. It should be remembered that data are processed not to hide defects but to enable interpretation, and to convince others of their validity. Interpretation itself can be regarded as working on two tiers: geophysical and archaeological. In turn, these aspects can either be treated either separately (normally with archaeological interpretation following the geophysical identification of anomalies) or simultaneously. Data must be verifiable, and the processing options discussed below will allow a geophysical interpretation of anomalies within data. However, a wider perspective will be required if a true archaeological interpretation is to be obtained. It is the archaeological interpretation that is the common and acceptable end point for all archaeological prospecting.

As discussed in chapter 4, a grid is usually established over the area in which a geophysical survey is to be undertaken, and this can be characterized by either local coordinates (e.g., eastings and northings) or through GPS-derived national grid coordinates. When data are collected randomly (for example, guided by a GPS instrument) they are normally re-gridded to a raster array with suitable spatial resolution in accordance with this grid. Alternatively, measurements can be made commensurate, creating a discrete raster grid of data points. The following discussion will focus on the computational treatment of such raster-related data.

For convenience of data collection and subsequent processing, survey areas are often subdivided into smaller blocks (e.g. 20 m × 20 m). These blocks will be referred to as data grids (DGs), although in much of the current literature DGs are often simply called grids, which is confusing and should be avoided. DGs are usually contiguous, and are meshed together at the end of a survey to form an overall data block, or composite. Once all processing of a final data block is complete, it is converted to an image (e.g., gray-scale) to be displayed or printed.

Using these concepts, it is convenient to split the steps that are commonly summarized as processing the data into three blocks: stage 1, restoration; stage 2, data processing; and stage 3, image enhancement. The procedures described here are largely categorized and labeled according to conventions used in the software package Geoplot 3 (Geoscan Research). Those using other commercial software will readily recognize the procedures, although they may be known by different terms. A number of alternative techniques have been described in the literature (Ciminale and Loddo 2001; Neubauer et al. 2001), that were also guided by the principle of achieving the best possible presentation of collected geophysical data for subsequent interpretation. Not all processing steps are mandatory, and for high-quality data only a few will be required.

Stage 1. Restoration

The steps undertaken during this preprocessing stage largely concern resolving minor errors associated with the type of instrument used and with issues regarding field procedures for data acquisition. In general, they maintain the value of each reading, although baseline shifts between lines of recorded data or blocks of data may mean the numerical value of a measurement may change. Most importantly, these procedures require information about the details of the data acquisition procedure (e.g., grid size, survey direction, number of sensors, etc.) that need to be documented carefully.

The most common corrective steps that are undertaken for magnetometer data during the restoration stage are termed here as de-sloping, zero-mean gridding, zero-mean traversing and de-staggering.

Slope Errors

Although most modern magnetometers are very stable, many instruments drift to some extent. Drift shows as a gradual change of the background reading over time, and the effect is particularly noticeable in fluxgate gradiometers. In data plots, it can be seen as a gradual change of background value for each subsequent survey line, leading to the overall appearance of a slope in the DG. The majority of fluxgate gradiometers can be set to record at the end of each DG the amount of drift over the zero point (log zero drift, see chapter 4). When data are downloaded to a computer, this value is automatically used to correct all measured data. The drift value is divided by the number of data points in the DG, and an appropriate correction is applied to each measurement. Only the corrected grid is saved, together with a record of the drift value. This procedure is only successful if the drift is linear over time, but unfortunately this is not always the case (Shell 1996). By subdividing the survey into smaller blocks (DGs) or collecting the data more quickly, this problem is reduced.

However, having to return to the zero point for drift recording can be time-consuming, and a software method that is equally effective is more often used. The drift in a DG can easily be estimated by comparing values along the first and last line. The difference between these values can be then applied to de-slope the affected DG (fig. 5.1).

Processing data with zero-mean traversing also removes gradual changes in the traverse background. If this step will be used later anyway, de-sloping may not be required. However, in all other cases de-sloping is preferred as it maintains the relationship between adjacent survey lines. Zero-mean traversing should not be used simply as a replacement for de-sloping.

Zero-Mean Gridding

This correction is required if the background reading of a survey shifts at some point, resulting in a fixed difference in each reading from one subset of data to another. The most common cause for a shift in the base

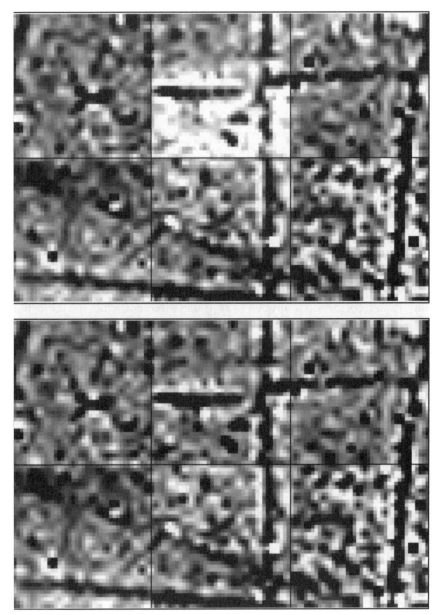

Figure 5.1. Slope correction. *Top*, untreated data; *bottom*, after slope correction of +3 nT. Data gray-scale for both images is −3 nT white to +3 nT black.

reading is when a large site is surveyed and the zero-reference position is moved to a more suitable point nearer to where data collection is taking place. In such a situation, it is difficult to ensure that the two zero values are equal. If they are not, in a gray-scale plot this will manifest itself as data blocks that are consistently lighter or darker than others. The problem can arise if different zero points are used, some being over areas of higher magnetic susceptibility. This is not an issue with most total-field instruments, since they do not need to be set to zero. On initial inspection, data collected by operators of different height may appear to be similarly affected, but in addition to a shift of the background value, the data will also be scaled, because varying distances to magnetic features will change the amplitude of anomalies.

To correct for a discontinuity, the mean value of a DG is calculated and then subtracted from each data point in it, which results in the mean of the new DG being zero. Evidently, if there are some extreme values from the presence of ferrous or fired material, for example, then the estimation of the mean should be constrained by rejecting any outlying values. The rejection threshold is usually determined either via statistical analysis or by some value chosen with respect to local conditions. An automatic method for rejecting outliers would be the use of the DG median for the correction.

This correction is often used as a quick fix to get a rapid overall picture of the data (see fig. 5.2). Many analysts use this as a first step before undertaking other processing, as some algorithms will be ineffective if the background is not close to zero.

However, the calculation of a mean may not be the perfect estimate for the background value of DGs, and applying the zero-mean gridding (ZMG) process can still result in each DG having a slight offset compared to its neighbors. In data with low amplitude, this residual effect may be undesirable for further processing or display. The best way to remove this artifact is by adjusting offset values so that data on either side of the edge between neighboring DGs have the same overall value. This is normally termed edge matching. This can either be done by choosing the edge of a particular grid and automatically adjusting the neighbor's offset, by using a software program slider bar to manually adjust any offsets, or by calculating offset values that minimize all edge mismatches in a data set (Haigh 1992).

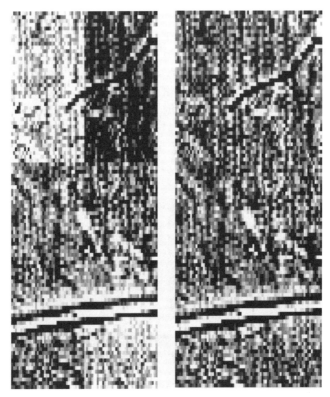

Figure 5.2.   Zero-mean gridding. *Left,* 100 m × 40 m fluxgate gradiometer data set shows typical checkerboard pattern created by a change of zero-reference points in the raw data; *right,* the image has been automatically corrected via the grid mean to give a clearer first view of the data, evidently with other minor defects uncorrected. Data gray-scale for both images is −3 nT white to +3 nT black.

Zero-Mean Traversing

Zero-mean traversing (ZMT) is a procedure to correct slight data baseline shifts between adjacent survey lines, which can be introduced as a result of collecting data in zigzag format or through the use of multi-sensor arrays. The former is caused by slight heading errors (i.e., variation in readings when the instrument is carried differently) and the latter by inherent differences between sensors. In a gray-scale plot, the effect results in stripes along survey lines, which become wider bands for sensor arrays. The heading error of fluxgate gradiometers results from often only slight inconsistencies during setup. When a poor zero point has been used in setting up, this artifact can be particularly obvious. A common error in establishing a zero point occurs when there is ferrous material in the vicinity of the chosen point. The setup will not be correct in this scenario, although the instrument may appear to be correct within acceptable tolerances. A large ferrous-free area around a zero point is evidently more critical when an array of fluxgate sensors is used.

It can normally be assumed that the mean value along each survey line should be roughly the same for all lines unless there are significant anomalies. To reduce the striping effect, this mean is calculated individually for each traverse and then subtracted from all data points along that line. This reduces the mean of the newly calculated data line to zero, which gave rise to the term zero-mean traversing. Since the mean of each traverse is zero, it follows that the whole DG has a mean of zero; zero-mean gridding is hence not required if zero-mean traversing is used in processing. As with the corresponding DG procedure, outliers and particularly large values have to be excluded from statistical evaluation. This can be done either by calculating the median of each line or by setting a cutoff that limits the range of data used for the calculation of the mean (see fig. 5.3). It should be noted that even well set-up instruments can sometimes benefit from applying ZMT, even if data have been collected in parallel format.

In an extreme case, where the zero reference has been poorly chosen, or the instrument physically knocked after setup, the overall survey raw data can appear incomprehensible. However, in some cases a zero-mean-traversing correction can minimize this defect (See fig. 5.4).

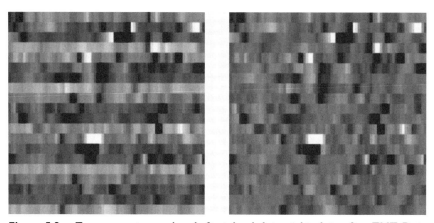

Figure 5.3.   Zero-mean traversing. *Left,* **striped data;** *right,* **data after ZMT. Data gray-scale for both images is −3 nT white to +3 nT black.**

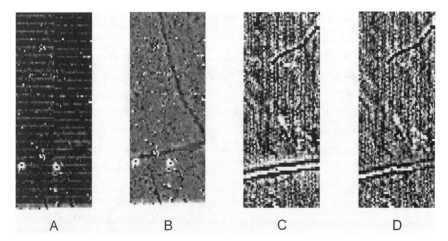

Figure 5.4. Zero-mean traversing with two 100 m × 40 m fluxgate surveys. Image A represents raw data, clearly in need of de-striping. Image B has been de-striped with ZMT, leaving visually cleaned-up raw data. Important note: anomalies that run along a traverse can be reduced relative to raw data when wrong cutoff levels are chosen. Image C data has been subjected to ZMT, altering some linear anomalies aligned with the traverse. This could lead to incorrect archaeological interpretation because image D suggests differing magnetic susceptibility levels along the traverse. In extreme cases where a linear anomaly stretches the length of a grid, ZMT can strip the anomaly away (see fig. 6.13). for all images is −3 nT white to +3 nT black.

While this correction is commonly used in the analysis of magnetometer data, there are a number of factors that should additionally be considered during the ZMT procedure.

Evidently, estimating the baseline level of a single line through calculation of an average will not always be correct, and it is probable that the correction will not eradicate the striping completely. Under these circumstances, the procedure may have to be repeated using new parameters for the rejection levels. In fact, it is almost impossible to get rid of striping completely, and many published examples show slight mismatches between lines of data. If the background level is not the same for adjacent lines, then this simple solution is likely to fail. This is particularly true when survey has been conducted next to, or over, a buried ferrous pipe, or is adjacent to a brick building.

Similarly, if a feature such as a ditch is aligned directly along a traverse, it may require a significant amount of trial and error to ensure that the anomaly from the feature is not reduced to zero and therefore de-striped from the data set. The only way to ensure that this does not happen is to review the data after ZMT with respect to the original data set. This can be either a simple visual inspection of two images or via software to map the differences between two data sets.

An additional benefit of this processing technique is that the knowledge of how it works can be important in the planning stage of a survey. For example, modern and ancient plowing can often create a regular pattern within a magnetic data set. This pattern is often hard to eradicate if it is at an angle to the direction of a walked traverse; if the traverse can be aligned along the line of the plowing then the effect from the plow can be significantly reduced with the use of zero mean traverse.

In addition to reducing the offset for each line, the ZMT function of some software packages also allows the removal of a linear trend along a line. This function can be useful, for example, if diurnal variations have to be removed from data that were recorded with single sensors, rather than gradiometers (Becker 1995). Figure 5.5 shows such data, collected with

Figure 5.5. Investigation of a Scythian settlement (Cicah, Siberia; see Becker and Fassbinder 1999) clearly showing grubenhäuser (pit houses), *north to top*. Surveyed with a Scintrex Smartmag SM4G-Special cesium magnetometer, dual-sensor configuration, 10 pT sensitivity, collection rate 10 Hz, 1 Hz band-pass filter; line spacing at 0.5 m; data resampled to 0.25 m in-line, ZMT-processed (with de-sloping), edge matching, and high-pass filtering; Data gray-scale is −5 nT white to +5 nT black. Short black stripes in NW grid are caused by heading errors over uneven terrain. Size of DGs is 20 m × 20 m. Image courtesy of Jörg Fassbinder and Helmut Becker.

two cesium magnetometers in duo-sensor configuration (i.e., two sensors in an array with 0.5 m horizontal separation, rather than in a vertical gradiometer). Measurements are affected by diurnal variations of the earth's magnetic field, but over the length of each DG (20 m × 20 m, in this example) the effect will be small enough to be corrected by removal of a linear trend (Tabbagh 2003).

De-Staggering

An unwelcome aspect to collecting magnetic data using a time basis for positional measurement is that it is often very difficult to walk at a constant speed. This is particularly the case when the data are collected in zigzag mode and the traverses are oriented over sloping ground. The position of the magnetometer in forward and return traverses can be shifted against each other. When plotted in the assumed middle position, however, the data can look quite alarming; at worst they can look like a zipper (see fig. 5.6). This phenomenon is most obvious when traverses are at or nearly at right angles to a linear anomaly.

There are a number of ways to minimize these errors in the field. Firstly, collecting the data in parallel mode usually reduces this problem, as it is normally easier to keep an accurate pace when walking downhill or uphill, rather than alternating. However, this does not guarantee perfectly positioned data; conditions underfoot or fatigue can cause variations in pace. Secondly, it is helpful to walk along marked lines; operators do not stray from a line, and have guide markers at fixed intervals, thereby recording data in the correct places. Thirdly, reducing the length of the traverses, or at least the distance between fiduciary (marker) points is helpful, as it is easier to keep to a constant walking speed over shorter distances.

However, even the best collected data may show elements of staggering, and these must be corrected before any processing is undertaken. Essentially, there are three ways commonly used to correct for this defect. First, the values on each adjacent traverse can be correlated and the traverses aligned automatically (Ciminale and Loddo 2001). Second, a manual correction can be used, shifting data along the lines by a certain number of data points. This correction can be the same for all traverses, or a single line can be corrected. The third method uses a speed-dependent shift value to automatically improve the data (Becker

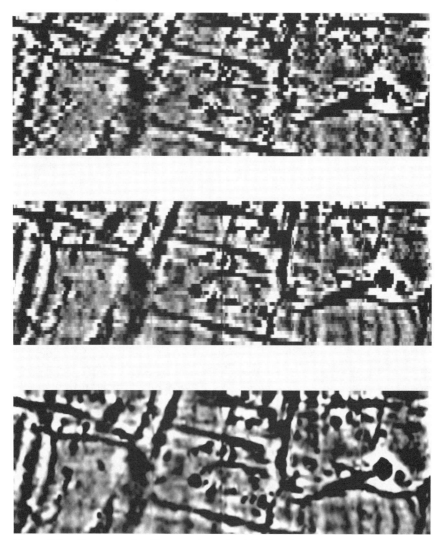

Figure 5.6. De-staggering. *Top*, raw data after ZMT; *middle*, data after de-staggering; *bottom*, interpolated sin(*x*)/*x*, for final image. Note that the vertical edges of the grid now show because of the dropouts in-filled by shifting values along the line. Data gray-scale for all images is −3 nT white to +3 nT black.

2001). If an array of sensors is used, the lines of linked sensors must be manipulated together. It is important always to check whether a correction has made the data better or worse. This decision is particularly difficult to make if anomalies are not linear, and especially if data contain closely packed small pit type responses.

An important fact to note regarding the way de-staggering correction works is that the position of each measurement is moved either left or right along a particular traverse. In doing so, a single end data point (or more, if the staggering is significant) will go beyond the limit of the traverse, and at the other end of the traverse the same number of positions will have to be created. The latter is a problem, as a judgment must be made as to how to fill the missing values. Depending on the data set, the blank values may be filled by taking an average of the values around the point, the traverse average, or even simply by assigning the value zero.

Stage 2. Data Processing

This second stage of analysis involves the use of algorithms that change individual readings with respect to other measurements, both on adjacent lines and along data traverses, irrespective of boundaries between DGs. This stage operates on the merged data block, the composite. The objective in this processing is to highlight anomalies that are within the data set, but are obscured by the variations in the background noise. It is in this part of the data analysis that the algorithms that are used can change the data in the hope that important archaeological information can be identified. This is different from image processing which is part of the visualization of magnetic data sets.

De-Spiking

Perhaps one of the most commonly used processing tools is to eliminate random spikes that appear within magnetometer data sets. In most magnetometer surveys, these anomalies (both positive and negative) are found to be the result of small items of ferrous material at or near the surface. There are also other factors, however, that may produce similar responses. Any small item of high susceptibility, such as a piece of igneous rock, can have a form and strength of response similar to that of an iron fragment, as we have seen in chapter 3. While all of these causes may be regarded as nonarchaeological, the response from an archaeological pit can look similar to a spike from iron if the sampling density is coarse by comparison to its size. In summary, this defect correction should be used sparingly because it cannot be assumed that all spikes are the result of modern ferrous material in the topsoil. Indeed, an archaeo-

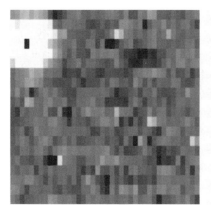 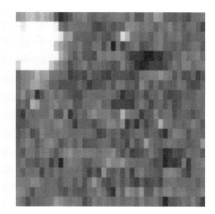

Figure 5.7.   De-spiking a 20 m × 20 m data block. *Left*, raw data; *right*, de-spiked data. Data gray-scale for both images is −3 nT white to +3 nT black. The majority of small spikes have been eradicated. Notice large negative anomaly that remains, however, after removing the central spike, *top left*.

logical site may be identified through an assemblage of spikes caused by the spread of nails.

The benefit of removing small spikes in data is that it is often easier visually to analyze data if these distracting anomalies are eradicated. The most common approach is to identify isolated high or low readings and replace them with some appropriate value. They are usually replaced by the average of the surrounding readings, however, and this often leads to misinterpretations: diffuse negative responses that usually surround positive spikes are left untouched by this process (fig. 5.7).

Routinely, some archaeological geophysicists de-spike in the preprocessing stage. By and large, that is not always a good idea because of the limitations outlined above. Additionally, it is very easy to exclude these spikes from the statistical analysis that is required, for example, for ZMT without getting rid of the actual value. However, there are some circumstances where de-spiking can be crucial, especially prior to filtering, and indeed there are occasions, such as the analysis of battlefields (see chapter 6), where the location of spikes can be highly relevant. Interpolation without de-spiking can produce distorted images (see below).

High-Pass Filtering

While de-spiking removes anomalies that vary on a very small scale, it is often also desirable to suppress data that vary gradually over a large scale. In particular, these may be anomalies caused by geological features or by a strong gradient in the variation of topsoil magnetic susceptibility. The processing to use in this case is called high-pass filtering because it allows only anomalies that are narrow (i.e., of short wavelength) to pass through. In the most common implementation, a fairly large area is selected around each data point, say 5 m × 5 m, in order to assess the background value around that point, for instance by forming the mean around the edges. This estimated background value is then subtracted from the central reading, and the process repeated for the next data position. The size of the analysis area is variable and determines how large the subtracted background is by comparison to the anomalies that are to be isolated; relevant software implementations allow variation of this important parameter. The main purpose of such processing is to highlight localized anomalies, often associated with archaeological features, and to suppress broad changes caused by other sources (see fig. 5.8).

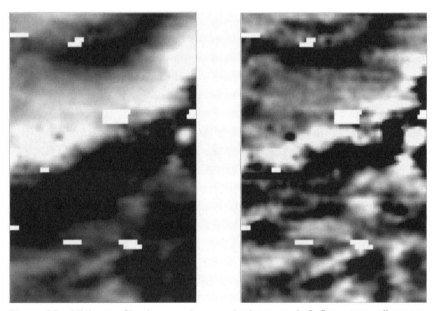

Figure 5.8   High-pass filtering, *north to top, both images. Left,* fluxgate gradiometer data from a temporary settlement site in Sri Lanka showing broad bipolar anomalies from relatively shallow geological sources. *Right,* after high-pass filtering, new anomalies, probably archaeological, become visible within the influence zone of the geological body (especially along the western edge). Filter strongly exaggerates negative halos created around some SE anomalies, however. Data gray-scale is −35 nT white to +48 nT black before HP, −15 nT to +16 nT after HP. Size: 20 m × 30 m.

However, the application of the filter can also create artifacts in the data. As with all other data processing, the removal of one data component usually comes at a cost, and with the high-pass filter it is the introduction of halos of negative data around steeply rising anomalies. A comparison with unfiltered data will usually allow the identification of areas where this unwanted effect is prominent, so that inappropriate interpretation can be avoided.

Periodic Filtering

Surveys undertaken with handheld magnetometers often show slight variations of signal strength along a line that are caused by height changes as the operator walks across the ground. If consistent, a periodic defect is introduced that can be linked to the gait of the surveyor. In many instances the consistency extends beyond individual lines; for example, if every line is started with the same foot, resulting data may show a banding *across* the lines, resulting from consistently higher and lower readings at particular positions.

Firstly, the period of the variations and the position of the first maximum have to be established, and this is usually done through spatial-frequency analysis. The results are then used in a second step to minimize the undulations. This periodic filter can only be used productively if the effect is very consistent, and is therefore not always successful. It is therefore important to minimize this effect during data acquisition by varying the height of the magnetometer as little as possible.

Activity Analysis

As discussed before, spikes in magnetometer data may have archaeological significance and can indicate areas of increased human activity. It is therefore useful to analyze the variability in a magnetic survey and to chart a suitable indicator across the whole site. One possibility is to calculate the variance of data values within a rectangular block around each data point (i.e., the sum of the squares of all deviations from the local mean) and plot the resulting values (see plate 4). Such a variance map can be used in two ways. If applied to the full range of magnetometer data (after appropriate data restoration) the variance will be greatest where narrow bipole anomalies result in rapid changes between positive and

negative data, and each will be replaced with a single value of high variance. This helps to "block together" several magnetic anomalies that show
individual high-low traces but can be interpreted as single entities (e.g.,
buildings). Alternatively, data may first be clipped to their positive values,
and the variance map then becomes more sensitive to the separation between neighboring anomalies.

Reducing to the Pole

Several processing techniques for magnetometer data have been developed that are mostly used in geological applications (e.g., analytical signal, Euler deconvolution, wavelet analysis, etc.). Most of these have been
tried on archaeological data but so far have found limited application in
routine archaeological processing tasks. Firstly, archaeological features often produce only weak magnetic anomalies, and collected data show high
levels of noise (see chapter 3), which is very problematic for most geological processing algorithms. Secondly, archaeological geophysics data are
usually collected with fairly high spatial resolution, and the underlying archaeological features can often be identified even with minimal processing. However, magnetometer data always show the typical signature of
bipolar anomalies, which are sometimes perceived to be confusing. As a
result, some authors have attempted to use inversion techniques to produce results that resemble the real shape of archaeological features (e.g.,
Eder-Hinterleitner et al. 1996). A somewhat simpler processing technique is reduction-to-the-pole (Blakely 1996). Since magnetometer results are "potential field data," all measurements are mathematically
equivalent and can be converted into each other (Tabbagh et al. 1997). For
example, a fluxgate gradiometer survey can be used to calculate results that
would be obtained with a single-sensor cesium magnetometer. One possible conversion is the calculation of data that would be measured if the
features were located at the magnetic pole. As discussed in chapter 3, all
anomalies at the magnetic pole would be symmetric, with the positive
peak centered over a feature and only a minor negative halo around the
perimeter. It could be considered that the shape of anomalies would then
relate more effectively to the underlying features. Although this has been
tested with some archaeological data sets (e.g., Schmidt 2001), many archaeological geophysicists find the negative troughs of typical anomalies

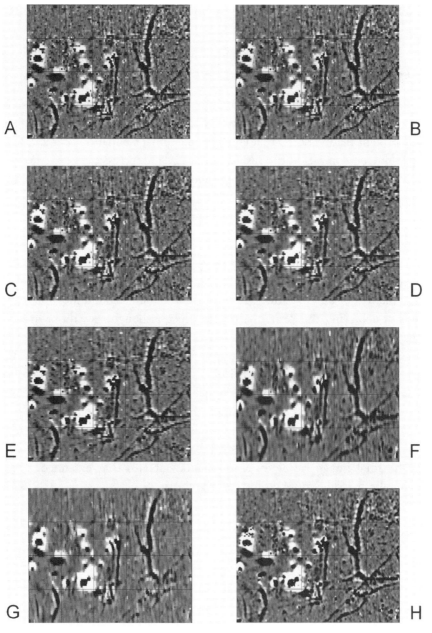

Figure 5.9.   Various smoothing options, with data gray-scale −3 nT white to +3 nT black. Data collected at 0.25 × 1 m intervals; (A) raw data; (B) low-pass window 1,1; (C) low-pass 2,1; (D) low-pass 2,2; (E) low-pass 3,1; (F) low-pass 4,1; (G) median filter1,1.The larger filters smooth the data too much, losing archaeological information. Image H (low-pass 2,1, interpolation × 2) shows the use of a low-pass filter for a finished diagram.

useful for data interpretation and do not require their removal for routine archaeological interpretation.

Stage 3. Image Enhancement

An eventual goal of all geophysical prospection is the production of images that are readily interpretable to archaeologists in order to rationalize subsequent fieldwork. To that end, adequate image enhancement is essential. Having completed the restoration and data processing steps, the following procedures can be selectively chosen to produce the most effective outcome of a survey.

Smooth Low-Pass Filtering

When working toward a final image, some analysts like to smooth the data. This is normally accomplished using a filter of some description, such as a median or low-pass filter. Both of these can eradicate minor spikes, as well as achieving the desired cosmetic effect, although they differ in how they identify the new value for each data point. In the case of a median filter, the reference value is the middle value within a sorted block of data, while the low-pass algorithm calculates the average within the block. If data are used correctly, then it is possible to reduce small-scale elements within the data, which can be distracting when viewing the final image.

Care must be taken when smoothing, as it is possible to lose narrow anomalies pointing to archaeological features. The bonus is that not only is the final image easier to comprehend, but also that extended, weak anomalies become more visible (see fig. 5.9).

Separating the Data

Evidently, as gradiometer data are processed such that values are distributed around zero (e.g., ZMG and ZMT), it is possible to separate the data into positive and negative values. At first sight, this seems to be an odd process to undertake on data when the anomalies essentially include both positive and negative values. Since a survey aims to convince others of the veracity of an interpretation, it is often useful to simplify the data. However, as we have seen in chapter 3, features tend to produce either a

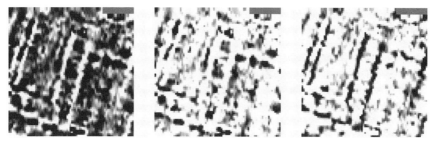

Figure 5.10. An illustration of the value of separating data into positive and negative values; negative anomalies show the presence of walls. Data gray-scale, *left,* −3 nT white to +3 nT black; *middle,* +0.1 nT white to +1 nT black; *right,* −0.1 nT white to −1 nT black. The smaller display range corresponds with reduction in data spread.

dominant positive or negative peak, and it is often this peak that allows identification of the anomaly. By separating data into positives and negatives, it can sometimes be beneficial in highlighting significant parts of anomalies, while reducing the visual noise associated with the smaller opposing responses that also constitute anomalies (see fig. 5.10).

Of course, there are some potential pitfalls in dividing data in this way. In particular, disassociation of the bipolar elements of some anomalies may lead to errors in interpretation. As an alternative to separating data into two groups, a visual equivalence may be obtained by using some of the display options noted below.

Interpolation

Interpolation is simply a tool used to visually enhance a data set by calculating additional data to increase spatial density (e.g., from 1 m to 0.5 m traverse spacing). This data expansion will smooth rough edges inherent in relatively coarse samples. Sometimes the data set is deliberately sparse, or is simply inadequate for the scale of the display that is required. In reality, nearly all magnetic data benefit from this enhancement because the process can be used to equalize the density of measurements in both directions. Equal sample densities were identified in chapter 4 as the ideal data collection strategy, and in part this is because square pixels are easy on the eye. However, a note of caution must be made here, as interpolation cannot increase the number of *real* data points. In fact, if interpolation is applied before the data are processed, then the time to process will be dramatically increased for no benefit. The advantage that can be gained relates to certain display options and visualization. In a GIS program, for example, it may be necessary for an analyst to view data at a very detailed scale, say 1:100. In such a scenario, it will be appropriate to interpolate, otherwise the image of the magnetic data will look too pixelated.

There are many ways to interpolate a data set. On data that have been collected on a regular grid, the options are usually either a linear interpolation between adjacent points or a nonlinear function such as sin(x)/x (Scollar et al. 1990). Given that magnetic data variations, by their nature, are nonlinear, the latter is usually preferred, although, as this is a cosmetic enhancement, the analyst may wish to experiment to find the best result. In some cases interpolation can smear rather than smooth, and clearly this is to be avoided (see fig. 5.11). Also it should be reiterated that no additional data can be produced by interpolation and, as it is chiefly a cosmetic enhancement, it is usually the last part of the process of analysis prior to final display.

Problems of Data Processing

As already noted in the discussion of individual data restoration steps, data acquisition errors can only be corrected if they are consistent. Subdividing an investigated area into smaller blocks (DGs) certainly helps, but it must be understood that processing tools cannot replace careful field practice. The adage "garbage in, garbage out" certainly applies to these algorithms as well.

Care has also to be taken when applying data processing to the complete assembly of survey data as nearly all processing techniques have side effects; for example, high-pass filtering introduces fringes around strongly varying anomalies (see above). Processed data therefore always should be compared to minimally treated results to ensure that in the final analysis processing artifacts are not interpreted as archaeological features. It is useful to be aware of the effects that are possibly introduced by different processing techniques. Data processing software therefore does not offer a single button labeled "Process All," and a black-box approach to data treatment should be avoided (Schmidt 2003).

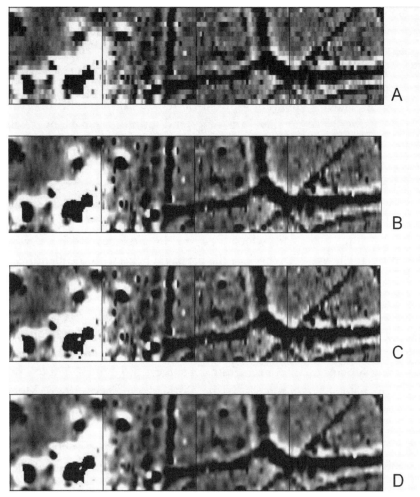

Figure 5.11. (A) Original data; (B) linear interpolation × 2; (C) de-spiked, low-pass filter 2,1, and linear interpolation × 2; (D) de-spiked, low-pass filter 2,1, and sin(x)/x interpolation × 2. All data gray-scales are −3 nT white to +3 nT black. Note variations and subtle changes among images.

Display and Visualization

Given that the objective of displaying magnetometer data is to facilitate their analysis and to convince others of the interpretation, individuals will have their own preferences as to how data should be displayed. There are, however, common attributes to a display that allow others to understand the image, and in some cases allow an analysis by another archaeological geophysicist without access to the original data. As computers have become more powerful and magnetic data sets larger, some of the display options described below have become less frequently used. That is not to say that some display options are now redundant, simply that they should be reserved for suitable applications. In broad terms, a display will either allow the viewer to comprehend the range and form of the data, or it will impose boundaries on the data in order to simplify the display.

X-Y Traces

This format is the most traditional of all display formats emanating from the study of analog magnetometer data. It facilitates the display of a full range of magnetic data, which is of great importance in their analysis. This format allows a first check of the data, and helps with the identification of unusual anomalies, or those that are the result of significant ferrous, fired, or igneous material. The measured points along each traverse (X) are joined together with the height (Y) of each point related to the data value at that point. This form of display was easy to produce on the smaller data sets that had to be graphed by hand, and some of the first recording systems used for magnetometer surveys produced X-Y traces on portable analog chart recorders (Clark and Haddon-Reece, 1973). More recent advances have enabled whole area surveys to be viewed by blocks of traces, which are often referred to as stacked profiles. In order to make large-area data or big data ranges more understandable, "hidden line removal" is often used to simplify the stacked traces. Even in the large digital data sets that are now routinely collected, X-Y traces are still important for the analysis of data, but are increasingly uncommon for final display delivery. If the objective of a display is definitively to identify very strong values, or for detailed analysis of individual anomalies when part of a data set is displayed at a suitable scale, then X-Y traces are a useful option (fig. 5.12). One of the shortcomings of X-Y traces relates to peaks of very strong anomalies, where the line from which the peak emanates is hard to identify. Also, where features run parallel to the X direction of the display, they are very difficult to see.

A variant of X-Y traces are 3-D displays, and they have only become possible with the digital analysis of data. Originally, this form of display was simply a perspective view of a stack of X-Y traces. Latterly, the deliv-

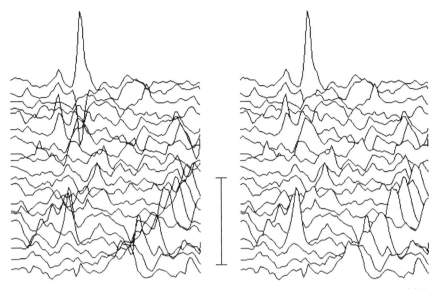

Figure 5.12. *Left,* **X-Y profiles;** *right,* **X-Y profile with hidden line removal, which simplifies images; vertical scale between images, 10 nT.**

ery has become more sophisticated, with the production usually taking the form of a surface fitted or modeled to the data (see fig. 5.13).

Contours

Again, this is a very traditional form of display, and involves linking data of equal value together, as with a topographic contour map. Well

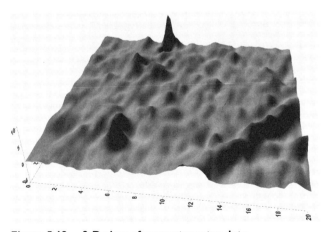

Figure 5.13.   3-D view of magnetometer data.

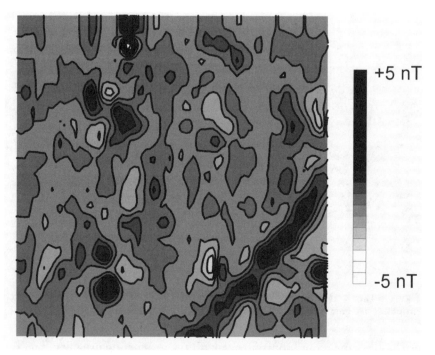

Figure 5.14.   Contoured data at intervals of 0.5 nT. To enhance the display, intervals have been filled using a gradation of gray levels as indicated.

chosen contour intervals can be very effective in forcing the eye to look at particular anomalies of interest. The intervals can be bound by simple lines or flooded with color. While some recent academic publications have used this form of display, contours in reality are only really suitable for small data sets. Contouring of large data sets that are typical in archaeological investigations is uncommon because fine details cannot be identified (Figure 5.14).

Dot-Density

This form of display requires analysts to choose minimum and maximum values (called cutoffs) of magnetic response that they wish to view. Each magnetic measurement is allocated a number of dots related to its value with respect to the cutoffs chosen; the larger the value designated the greater the number of dots. The dots are randomly positioned within

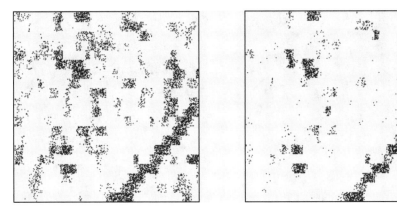

Figure 5.15. *Left,* dot-density plot, data scale from 0.1nT white to 3 nT black; *right,* the same cutoff levels, with positive contrast applied to the data to highlight higher values.

an area, normally rectangular, that represents the position of the reading within the survey grid. A result is that a plan of the response can be produced where, normally, a value below the minimum cutoff will be white and a value above the maximum will be black. For gradiometer data, it is usual to have the minimum cutoff at or near zero, while the maximum is a positive value of a few nanoteslas (see fig. 5.15). Of course, a reverse scale can be chosen to show negative intensity.

The strength of this display format is that the eye focuses only on a specific range of data. As a result a dot-density map can be very helpful in convincing a third party of the interpretation of the data. The limitations are that this is a relatively crude display, in that it is neither possible to interrogate the whole of the data set with a single map nor to understand the form of a particular response. While in the past this has been a commonly used display format for producing flat representations of data that could be placed directly onto a map of an area, it has largely been replaced by gray-scale diagrams.

Gray-Scale

Gray-scale images are by far the most frequently used display format for magnetometer data collected for archaeological purposes. Cutoffs are chosen in a similar way to those for dot-density plots, but in this case each

139

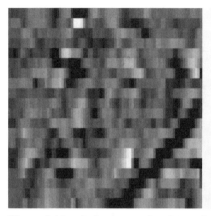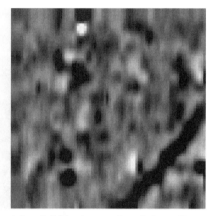

Figure 5.16. *Left,* gray-scale of raw data; *right,* sin(*x*)/*x,* interpolated data; Gray-scale for both images is −3 nT white to +3 nT black.

measurement is linked to a shade of gray that fills the associated rectangular box (see fig. 5.16). Depending on the nature of the data, the number of palette gradations varies from about 10 up to 255. While many analysts consistently use black to indicate the maximum cutoff, that choice is not universal, and some prefer to reverse this scale. As well, it is possible to color the palette, although this is less commonly done than in the display of other geophysical data.

GIS and Enhanced Visualization

The data processing functions and displays discussed in this chapter are incorporated in most software packages that are specifically written for archaeological geophysics data. Data restoration functions that are based on individual DGs are not normally found in nonspecialized programs. However, once all processing has been completed, it is often desirable to incorporate the results into a presentation or analysis package, CAD or GIS, for example, to produce interpretation overlays. It should always be possible to export data from a processing package as a data image (e.g., a gray-scale or color diagram) and then to import this picture into the presentation software. If the site grid that was used for the geophysical data was in a different alignment than the orientation of the presentation diagram, an image rotation is usually required. Depending on the spatial resolution of the data image and the algorithm used for calculating the rotation, this may lead to artifacts in the final presentation diagram. As a result, it is rec-ommended to export data images with high spatial resolution.

The process of fitting the geophysical data image to its correct place in a presentation diagram is normally referred to as georeferencing (Schmidt 2002). When survey grids are set out with reference to an imag-inary horizontal surface (see chapter 4), like other map products, georef-erencing can normally be achieved with simple rotating, shifting, and stretching. Mathematically, these manipulations are referred to as affine transformations. Several mechanisms exist to store the required informa-tion in addition to the data images themselves (e.g, as "world files") or in geographic-aware image formats (e.g., GeoTIFF). However, if a survey grid was distorted by, for example, strongly undulating topography, more sophisticated local stretching of the data image may be required.

Only very limited manipulations can usually be done with imported data images, as the underlying data have been converted to a small num-ber of gray (or color) levels. For example, if contrast stretching or his-togram equalization are employed, the results will be less useful compared to an application of such techniques to the full data range. GIS software therefore allows users to import processed geophysical measurements as data rasters and to apply further processing and adaptive-display options. Unfortunately, the exchange of data between different software packages is not always straightforward, and several intermediate data rasters may have to be created.

Most presentation, CAD, and GIS software packages allow data to be imported or data images as separate layers that can be displayed on top of each other and switched on and off as required. This allows the spatial com-parison of anomalies in different data sets that may have been produced by the same buried archaeological features. Alternatively, results from two or more data sources may be combined so as to produce a composite image that shows their respective contributions (Kvamme 2006; Schmidt 2001). The simplest method is to add the stretched values of the data sets together and hence to create an image that shows features irrespective of their origin. However, care has to be taken not to eliminate anomalies in this process. A buried ditch, for example, will on most sites produce a positive magne-tometer anomaly (peak) but a low earth-resistance anomaly (trough). One of the data sets may have to be inverted to make the addition of the two results meaningful. The use of two-dimensional color palettes is another possibility to combine two data sources. Magnetometer data may be displayed, for example, with a color scale from blue (negative) to red (positive), and earth-resistance data with a scale from green (low) to yellow (high). Any combination of data values is then assigned a composite color (e.g., positive magnetic and high earth-resistance is red and yellow, represented as orange) and after some training an interpreter can recognize the spatial relationships of related anomalies. Alternatively, other display methods (see above) that can be distinguished by the interpreter can be used to represent different data. For example, earth-resistance data can be plotted as contour diagrams over a gray-scale diagram of magnetometer anomalies. Alternatively a 3D display can be generated from earth-resistance data, over which the gray-scale data of the magnetometer would be draped. These different distinguishable display methods are referred to as "visual classes" (Schmidt 2001).

ARCHAEOLOGY AND DATA INTERPRETATION

The preceding chapters have charted the progression of theory and practice in magnetometry. Here we will describe five broad areas of archaeological application. The intention is not to show the typical or best responses for all site types, as it is felt that images and case studies can become repetitive. Rather, the aim is to look at five broad topics that are familiar to most archaeologists, and within them analyze the complexity and limits of interpretation of case studies. The topics cover settlement, industrial, garden, and ritual and religious sites, and anomalies resulting from igneous and ferrous material.

The data from the magnetometer surveys will be interpreted with reference to the theory and processing that has been outlined earlier, and will draw, where appropriate, on additional information from other geophysical techniques and excavations. It is important to understand that there is a significant difference between an archaeological and a geophysical interpretation of magnetometer data; the former is considerably more subtle, and requires an understanding of the context of the anomaly as well as knowledge of the site. In short, a holistic approach is recommended, as it is not productive to consider magnetometer measurements in isolation. To date, the most abundant data have come from the use of vertical fluxgate gradiometers, and the case studies in this chapter reflect this situation. However, it should be stressed that any of the other magnetometer types could have been used to investigate these archaeological problems.

Settlement Sites

One of the most common uses of magnetometer surveys is to map and delimit zones of former settlement, especially of periods or cultures that maintained permanent or semipermanent places of habitation. The size and complexity of settlements can vary tremendously depending on factors such as the number of inhabitants, the nature of occupation and the longevity of the site. Frequently, the data are dominated by key archaeological elements, such as enclosure ditches. Even when these are not present, it is likely that other patterns of human origin will be apparent in a settlement, such as pits associated with storage or refuse disposal. Individual houses or dwellings are sometimes located in the data, but often the structures associated with settlements are difficult to detect because of the small size of features such as postholes. Where occupation was short-lived, interpretation of the associated magnetic anomalies is likely to be much easier because of simpler archaeology. Short-lived sites do, however, have some unhelpful aspects in that the strength of anomalies may be weak from a lack of enhanced magnetic material in the fill of ditches and pits. Additionally, some short-lived occupation sites, especially those linked to nomadic peoples, leave little tangible evidence that can be found using geophysical techniques. The following examples have been chosen to represent common or recurring elements of different types of settlements, but they are not exhaustive.

Small-Scale Village Settlements

The image in figure 6.1 shows typical magnetic responses from a small settlement site of late prehistoric or Roman date. There are clearly ditches defining the core of the site, which extends beyond the limits of the survey. The strongest responses are largely associated with the center or core occupation, and are the product of fires and burning that produce an accumulation of high-susceptibility material in ditches and pits. As such, the pattern is related to not only the density of occupation but also the disposal of rubbish. It can be seen that some of the ditches become weaker with distance from the center of the site. This can be observed on many sites, and is referred to as the habitation effect (Gaffney et al. 2002).

The responses at this site are fairly common for low-status settlements (see Gaffney and Gater 2003, or recent issues of Archaeological

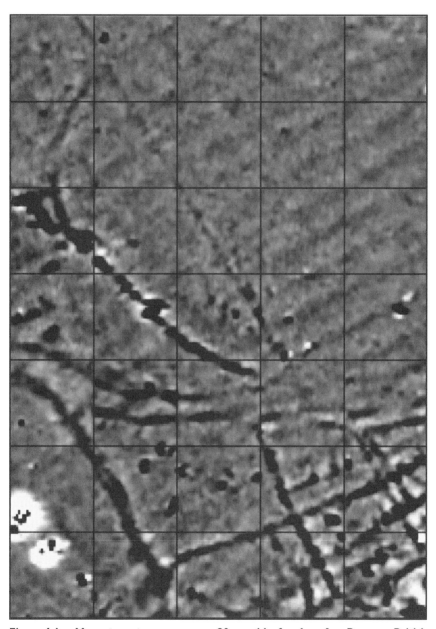

Figure 6.1. Magnetometer survey, on a 20 m grid, of an Iron Age Romano-British enclosed settlement, *north to top*. Note general reduction in anomaly strength toward north that reflects reduction in occupation debris with distance from the settlement southeast. Weak but regular patterning at 45 degrees indicates later ridge-and-furrow cultivation. Data gray-scale is −3 nT white to +3 nT black. Survey by GSB Prospection Ltd.

Prospection, for more examples). However, the interpretation may be complicated by elements of plowing (both recent and ancient) and modern large-scale ferrous responses. Fortunately, in this case, although both were present, neither inhibits the analysis of the data.

An example of an unenclosed settlement is Brown's Bottom, situated on the floodplain of the Scioto River, about 16 kilometers south of Chillicothe, Ohio, and belonging to the Hopewell Native American culture (about 100 BC to AD 400). Archaeological investigations at the site were prompted by discovery of surface artifacts in the 1960s. However, small-scale excavations revealed little, and the site was not given much credence until recent investigations were carried out using magnetometer surveys employing a structured research design. Field walking, geophysical investigation, auger survey and ultimately excavation provided an increasingly certain interpretation at Brown's Bottom. Although these were employed by a field school for university students, the strategy was one that had been developed by Jarrod Burks for commercial investigations.

It was to be expected that the sandy soils at the site would produce relatively low magnetic responses because of a low contrast between feature fills and the surrounding soil. The type of potential archaeological features at the site, those of unenclosed settlement, and the low magnetic contrast suggested that a dense mesh of data capture would be required. The survey was conducted using a Geoscan Research FM36 fluxgate gradiometer, with data collected at 0.125 m sampling along transects separated by 0.5 m.

The image in plate 5 was captured after minimal processing of ZMT, low-pass filtering and interpolation. The project leader then interpreted the anomalies via a classification based on signal strength. Despite the weak signal from the majority of the observed magnetic anomalies, there were 27 anomalies in the range 2–24 nT that were evidently not modern features or a result of ferrous debris. Looking at the image, it is apparent that pronounced negative values are associated with the strong positive anomalies, and it is reasonable to suppose that together they indicate areas of *in situ* burning.

In an effort to verify the anthropogenic nature of the anomalies, Burks used a soil auger to investigate many of the presumed cultural features. The majority showed clear signs of cultural fill (charcoal, heated and darkened earth, and an abundance of ceramics) associated with pits and ovens that are common at Hopewell settlement sites. The magnetometer classification and the auguring allowed the formulation of an excavation strategy that fulfilled the research objectives of that year's excavation, which was to identify features with subsistence evidence.

On excavation, the topsoil (plow zone) was found to be about 0.3 m thick, beneath which many archaeological features were found. The strong anomalies were correlated with ovens up to about 1.5 m in diameter and extending a further meter below the topsoil. The weaker anomalies that were identified also proved to be archaeologically significant with those of about 2–4 nT resulting from typical settlement features, such as less substantial pits. It was noted that the latter were much harder to discern during the excavation, and it is possible that many may have been missed in the absence of the magnetic data.

High-Status Settlements

High-status settlements have often resulted in exemplary responses to a number of geophysical techniques, especially electrical resistance. The reason why earth-resistance measurements prove so effective here is that they often contain more substantial structures than lower-status sites; frequently buildings constructed from stone or brick are major or significant elements of such settlements. Their construction indicates permanency and suggests, at least in parts of the world where stone is difficult to obtain, an investment of wealth. In the example of a Roman villa shown in figure 6.2, parts of a rectangular ditched enclosure can be seen. In this image the plotting scale of the magnetometer survey has been reversed from the majority of the images in this book; the positive anomalies are shown in lighter shades to give a direct visual comparison with earth-resistance data. Within the enclosure is a series of weak, negative, linear magnetic anomalies that result from the use of less-magnetic material in the construction of a building. This is a reasonably unambiguous magnetometer response for such a structure, and can be compared with the earth-resistance data that cover a subset of the magnetic area. The general results are compatible, although in some areas, such as the western edge of the building, the magnetometer data are sharper. There is no reason why the data sets should be identical, given the different phenomena being measured. The general spread of high-resistance anomalies within the building (indicating either intact floors or demolition material) is not

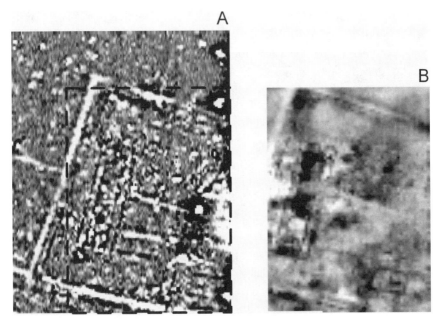

Figure 6.2. Integrated geophysical survey over a high-status enclosed settlement, *north to top. Left,* magnetometer survey data at 1 × 0.25 m density. Data gray-scale is −3 nT black to 3 nT white. *Right,* a resistance survey of the boxed area shown in the magnetometer image. Survey by GSB Prospection Ltd.

revealed in the magnetic data to any great extent. One must assume that the resistive body is magnetically fairly homogenous, as there are no coherent signals from the magnetometer supporting the interpretation that they represent constructed floors. rather than a magnetically inhomogeneous spread of demolition material.

At sites such as this, it is often the case that the nature of the magnetic response can be difficult to identify, for example when a weak negative response is the principal signal, as opposed to being the minor (negative) part of a stronger positive signal. A case in point can be seen in this magnetic image, where a well-defined narrow negative anomaly is located just inside, and parallel to, the western ditch. It is likely that the negative response does not form part of the ditch anomaly, given the distance between them. However, the negative data may be linked to the ditch, perhaps a buildup of nonmagnetic material, a bank or even a wall. Close inspection of the positive ditch anomaly reveals very slight negative responses on both flanks. Both of these form characteristic parts of the shape of the magnetometer anomaly, as predicted for a north-south aligned ditch.

From anecdotal evidence, it is often the case that analysts fail to identify stone structures in magnetometer data. There are a number of reasons for this, including absent, or very small, magnetic contrast with surrounding soil and a high "noisy" magnetic background that swamps the weak signal. In the latter case, highly contrasting changes in magnetic susceptibility can be a result of natural soil variation, modern factors such as dumped or made-up ground, or indeed archaeologically produced by random demolition material or archaeological strata randomized by the plow. Even in the case where individual stone walls cannot be identified, however, there may be some relevance in analyzing the background if the structures produce a coherent increase in noise from an apparent random configuration of fired and burnt material (see Gaffney and Gater 2003, fig. 77). In this way, the limit of any structure can be estimated and the detail can be confirmed by other geophysical techniques or by excavation.

Ancient Towns

The magnetic survey of ancient towns has a long history. One of the largest early surveys was undertaken to establish the location of the ancient city of Sybaris, situated in southern Italy. In her pioneering 1960s work, Elizabeth Ralph used cesium-vapor instruments in single-sensor mode to identify the location of the former Greek colony. She was well aware that she would map geological variations in this mode, but she subsequently showed that elements of the town could be identified even under depths of 4 m of alluvium (Bevan 1995).

In fact, many surveys over large occupation sites in this geographical area have often produced results of great clarity. Linington, who was a pivotal worker in the 1960s, illustrated that much of this definition was a result of the "red" soil of the Mediterranean area. In these circumstances, individual walls are often visible as negative signals within the highly magnetically susceptible, but homogeneous, red soil. On sites that lack naturally enhanced red soils, it is still possible to locate individual walls, although this is usually from anthropogenic enhancement of soil surrounding the walls. The origin of the surrounding material can be diverse, and includes general occupation detritus built up against the walls, deliberate consolidation, floors packed with highly magnetic material, and foundations cut through earlier deposits, often middens.

A landmark magnetic survey of the later part of twentieth century was the coverage of the Roman city of *Viroconium* (Wroxeter), which in the late second century was the fourth largest town in Britain. Of the 190 acres (78 ha) of land that lay within the defenses of the city, some 175 (70 ha) were available for investigation and subject to magnetometer surveys using Geoscan Research FM fluxgate gradiometers (Gaffney et al. 2000). The soils in the locality have often proved to be inherently poor at producing strong measurable magnetic signals from even relatively long-lived occupation sites. At a Roman villa near the city, the only significant magnetic anomaly proved, on excavation, to be the stoke hole associated with the villa's under-floor heating; the rest of the site was effectively invisible. A subsequent survey of the city proved to be a complete contrast to such surveys from its hinterland. The question is, why are anomalies within the city so different in strength from those on sites in the near vicinity? The answer to this question lies in the soils that can be encountered within this city. They are radically different in that they had been subjected to significant alterations over three centuries of intensive occupation. As a result, the archaeological features are embedded within a highly susceptible background made up of occupation debris, including burnt and fired material.

While the strength of the anomalies means that they are easily discernible using standard magnetometer systems, there are a number of matters of interpretation that should be considered. In this part of the data set (fig. 6.3), a host of issues arise: for example, towns are multi-period and often have a long continuous occupation. While an earlier defensive structure sweeps across the top of the image, there is evidence for a regular grid of roads and buildings within subdivisions of the city. The dominant aspect of this image illustrates that the most magnetically visible elements often reflect the latest, or grandest, use of the site. Even then there are differing responses from apparently similar archaeological features. This is particularly clear in the case of structural features that have produced anomalies, some negative and some positive. The former are routinely found at Wroxeter as the common building stone there exhibits weak magnetic properties in comparison with the highly altered and very mag-

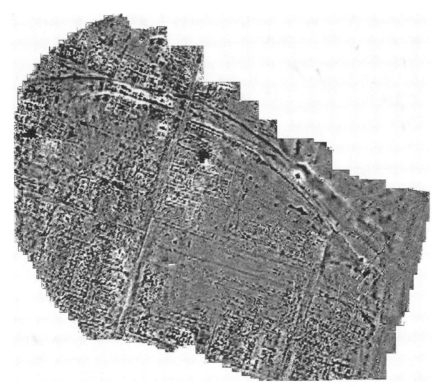

Figure 6.3.   A 22-acre (9 ha) image extract from magnetometer data collected at the Roman city of Wroxeter, Shropshire, *north to top*. Across the city there is a wealth of anomalies from structural elements, plumbing, industrial activity, roads, and the like. Data gray-scale is +3 nT white to −3 nT black. Image courtesy of GSB Prospection Ltd.

netic soil within the city. It is possible that two different sources of build-ing material were used in the construction of the urban houses. The in-vestigators at the site, however, believed that the difference in response arose from intense burning that magnetically enhanced the surface of the stone. Ordinarily, brick or construction material derived from an igneous source is unlikely to produce coherent low-positive responses when as-sembled as a wall. This is because the material was initially not *in situ*, and will produce an irregular high-low pattern that results from the addition and subtraction of remanent magnetization associated with the locale of its fabrication as a building material (Bevan 1994; Hesse et al. 1997). In-terestingly, at Wroxeter there is historic evidence for a major conflagration on the site (White and Barker 1998), and it is hypothesized that this fire scorched the surface of the stone *in situ*, thereby producing stone with a coherent and recognizable positive response. If this is true, then the fire can be mapped, at least in those areas where stone structures existed; the poorer areas, where wooden building material was more common, remain, as is so often the case, relatively invisible.

Another facet to be considered is that the background magnetic response is often high in towns because of the complex nature of the soils, earth strata, and the significant reuse of the site. The variation at Wroxeter is great, with some exceptionally strong anomalies associated with both industrial and settlement areas. Despite the high magnetic background resulting from the intensive use of the site, there is a central area that lacks the strongly contrasting anomalies seen elsewhere. In this case, it is possible that this area had been a cattle market (White and Barker 1998, 92). There are weakly magnetic subdivisions within this zone, but the archaeological interpretation is not certain. With these observations in mind, it should be remarked that no matter how clear the images obtained from magnetic surveys within ancient towns, the minutiae of interpretation cannot be as definite as that for a survey over a single-period occupation site.

One aspect often overlooked when towns are surveyed is the sheer density of archaeologically significant measurements. As a result, it is possible to go beyond the concept of wall-following and produce an archaeological interpretation that identifies potential use patterns within the site. At Wroxeter, for example, experience of other surveys, and knowledge of the likely spatial variation within a Roman town, have allowed specific zones to be identified. These include industrial, elite, and artisan areas, and examples of the data from these zones can be seen in plate 6. Analysis of the statistical nature of the magnetic background in each area indicates significant variation. It is possible that this variation may be used as a template to analyze other areas of the town where the archaeological interpretation of the anomalies is not as clear.

Another significant approach to the analysis of magnetometer data has been published recently by Benech (2007), who has studied the city plan identified by cesium-vapor data in terms of urban space. Specifically, he has looked at the regular subdivisions (the Hippodamian plan) of the town of Doura-Europos, Syria, and analyzed the variation of room sizes

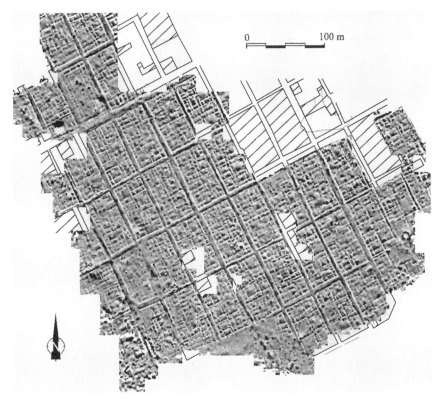

Figure 6.4. Magnetometer survey of the southern part of Doura-Europos, Syria, using a cesium-vapor gradiometer recording measurements every 0.2 m along profiles 1 m apart. Data gray-scale is −10 white to +10 black nT m⁻¹. Image courtesy of Christophe Benech.

within different blocks. This study is significant archaeologically, as it complements traditional avenues of investigation while bringing the uniqueness of magnetometer data to the fore (see fig. 6.4). In his analysis, he illustrates that the study of the organization of public and private space of a city can be augmented by magnetometer data in a way that pushes at the boundaries of archaeological interpretation.

Benech has significantly extended the interpretation of data beyond identification of dwelling units toward an analysis of room function, room size, and the articulation of space within the overall unit (fig. 6.5). As a result of both descriptive and statistical analysis of geophysical and excavated information (on, for example, the function and size of rooms) the interpretation of the data has been taken to a new level: the organization

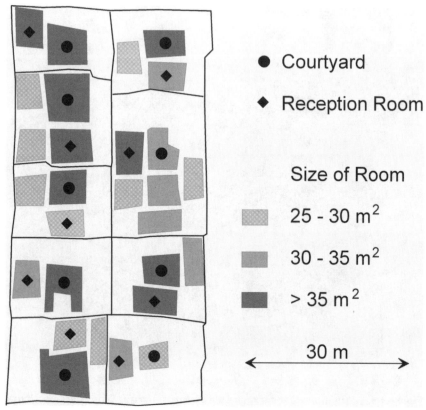

• Courtyard

◆ Reception Room

Size of Room

▨ 25 - 30 m^2

▧ 30 - 35 m^2

■ > 35 m^2

←——— 30 m ———→

Figure 6.5. Summary analysis of hypothetical divisions of dwelling units within block M2 at Doura-Europos, Syria. Image courtesy of Christophe Benech.

of public and private space in a city. This citywide view of the symbiotic relationship of social and cultural use of space reveals a use of magnetometer data that is considerably beyond wall-following.

Key Points

Occupation sites often produce strong magnetometer response resulting from the presence of soils and deposits that have been significantly altered by the settlement activities.

While multi-period occupation sites frequently produce clear magnetometer responses, these results may constitute a significant oversimplification with respect to the buried archaeological features of such sites. Later-period sites often dominate the data.

On many occupation sites, signals are strong enough to be mapped using all types of modern equipment. Choice of instrument and methodology is often dictated by the presumed size of the occupation site. Cart or sledge systems are evidently of value when a high-density measurement strategy is required.

Magnetometer surveys within ancient towns are often data-rich, both magnetically and archaeologically. There is often scope to analyze and interpret the magnetic measurements beyond the identification of individual buildings.

Industrial Sites

The label, industrial site, is often assigned to surveys that encounter very strong magnetic anomalies. These alone are, however, not sufficient criteria to classify such sites, as will be seen from the discussion below of igneous and ferrous anomalies. For the purpose of this chapter, industrial sites will be considered to be sites that were used for producing goods on a scale beyond domestic use. Most production activities require the use of heat, applied in kilns (e.g., for pottery, brick, tiles), furnaces (e.g., for metalworking, glass production), or hearths (e.g., for blacksmithing). Where these heating events were sufficiently substantial, they can lead to strong thermomagnetic anomalies that are easily detected in a magnetometer survey (Jones 2001). Most of these sites are characterized by ancillary structures that are essential to unravel the activities that took place at them in the past. However, the magnetic anomalies associated with these are often weak, and overshadowed by the strong thermoremanent signals of the industrial features. A similar problem is encountered in industrial buildings of more recent times, where ferrous construction materials were used, creating substantial bipolar anomalies that mask other structures.

Medieval Tile Kiln, High Cayton, North Yorkshire

The site of High Cayton was once a grange (i.e., a farm) of Fountains Abbey, a large Cistercian monastery in North Yorkshire. From 1135 until the Dissolution of the Greater Monasteries Act in 1539, this grange provided the abbey with a variety of goods, mainly farm produce (e.g., fish and wool), but also floor tiles produced in local kilns. These kilns were built on a southwest-facing slope to harness the natural draft, or flow of air, but have not left any visible traces on the surface. The results of a fluxgate gradiometer (Geoscan Research FM36) and earth-resistance survey over this area are shown in figure 6.6, where two of the kilns are clearly visible. Mag-

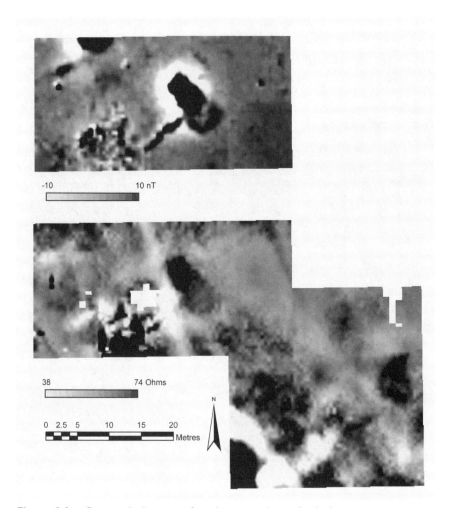

-10   10 nT

38   74 Ohms

0 2.5 5   10   15   20
Metres

N

Figure 6.6. Gray-scale images of an integrated geophysical survey over a medieval tile kiln site at High Cayton, North Yorkshire. Magnetometer survey, *top*, shows details of one kiln structure, confirmed and extended by an earth-resistance survey, *bottom*.

netometer data were recorded with a spatial resolution of 0.5 m × 0.5 m, and the processing comprised only automatic grid-balancing and spike removal. To improve their visual appearance, the data were bilinearly interpolated to 0.125 m. Earth-resistance data were recorded with the same spatial resolution, but their strong variation over the topographic slope led to a Geoplot high-pass filter being applied (radius 5 m), highlighting isolated features, even within areas of weak contrast.

The magnetometer data clearly show the layout of one full kiln, aligned along the contours of the slope, and parts of a second kiln to the northwest (at the edge of the grid). The main kiln has a rectangular body (ca. 4.5 m × 7 m) of high magnetic response (up to 120 nT), and is accompanied by a halo of negative data (ca. −25 nT). To the southeast is a rectangular area defined by a weaker positive anomaly. The latter is interpreted as the location from which the kiln was fired (the Stokehole). This composite layout fits well with published medieval kiln typologies (Musty 1974), and the data can be archaeologically interpreted with confidence. The data also show a linear anomaly extending southwest, exactly down the slope, and it is very likely that it relates to a structure built to create an updraft of air through the kiln. The data of the main rectangular anomaly appear fairly uniform in the gray-scale display, but inspection of the X-Y traces and 3D visualization (fig. 6.7) clearly show the different magnetic response along the length of the kiln. The higher response in the southeast is further evidence that this was the main firing area, whereas the northwest part was

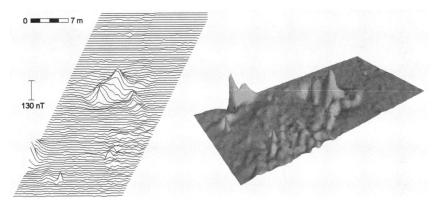

Figure 6.7. Alternative presentations of the magnetic data of Fig. 6.6, showing details of the High Cayton's kiln structure.

probably mainly used for stacking up tiles to be fired. This is supported by the earth-resistance data, which have higher values in the area of the magnetic anomaly, and especially in the northwest, possibly pointing toward a stack of material there. The elevated high resistance values to the southeast of the kiln show rectilinear anomalies, characteristic for structural remains, as would be expected in fuel-storage areas around such kilns.

West Bretton Blast Furnace, West Yorkshire

Incorporated into the Yorkshire Sculpture Park lie the remains of the West Bretton blast furnace. Historical evidence shows that it was established in 1720, and operated in conjunction with Kilnhurst forge until 1806. A furnace in nearby Rockley was also part of the enterprise. No upstanding remains are visible on the site, and a magnetometer survey was therefore used to understand better the layout of the West Bretton blast furnace. A leat (man-made water channel) entering and leaving the site and a slag dump are the only pronounced earthwork features now apparent. Data were collected with a 1 m × 1 m resolution, using a Geoscan Research FM36 fluxgate gradiometer. Some very high data values exceeded the instrument's recording range and were automatically registered as missing (i.e., dummy). To produce a more appropriate data representation, these missing values were replaced with estimates calculated with a Gaussian low-pass filter, a process similar to the stitching procedure described by Scollar et al. (1971). The data were then bilinearly interpolated to 0.125 m spacing, and the resulting image can be seen in figure 6.8. This interpolated representation

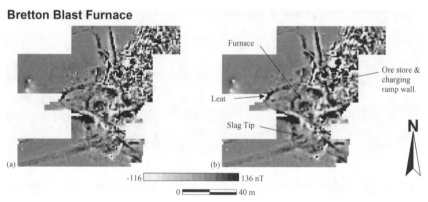

Figure 6.8. Magnetometer survey of a late-nineteenth-century blast furnace site at West Bretton, West Yorkshire. *Left,* a gray-scale image is derived from the original data; *right,* the furnace facility after data interpolation to enhance details.

makes the layout of the site easier to understand, masking the relatively coarse data resolution. Since some features produced much weaker anomalies than others did, it was necessary in the analysis to display the data with different clipping ranges so that the shape of weaker anomalies could be more clearly identified. The close proximity of these features left high-pass filtering unable to resolve their details. The survey was able to identify the major site components, including the furnace, an octagonal building housing it, and storage areas for charcoal and iron ore (Vernon et al. 2002).

Key Points

Industrial sites are usually characterized by several strong anomalies produced by an assemblage of features linked to production on an industrial scale. Strong bipolar anomalies often appear to mask weaker signals from associated features.

It is worthwhile to analyze the data carefully and in detail to detect structure within the strong anomalies and to verify weaker anomalies related to associated buildings, thus helping to interpret such sites comprehensively (Vernon et al. 1998). Displaying data with different clipping ranges can often improve such evaluations.

Ritual and Religious Sites

An important area of archaeology is the investigation of belief systems, especially with respect to religion and death. These aspects of former societies are not well understood, partly because the sacred sites are at one extreme based on organic materials, such as trees, and at the other on substantial stone edifices. Some of the sites appear to have been only used intermittently, or at specific times of the year. The exact nature of the ceremonies at such sites can often only be explained via speculation, and they are frequently given the catchall term of ritual sites.

Given the fact that these enigmatic sites contain neither settlement nor industrial activity, and that any activity may have been confined to specific times of the year, it would be unlikely that they produced significantly enhanced magnetic material. A cursory glance at English Heritage's web-based database, for example, (http://sdb2.eng-h.gov.uk/) will confirm that many early ritual sites (e.g., henge monuments and causewayed enclosures) have produced few anomalies of interest. However, many of

CHAPTER SIX

these surveys were undertaken with older instruments and it is possible that both survey methodology and instrument insensitivity may be important factors in the lack of mapped responses. By way of contrast, some modern magnetometer surveys have produced data images that have revealed highly significant details.

Prehistoric Ritual Sites

There are many prehistoric sites whose former use cannot be universally agreed upon. These are often called ritual, but it is clear that they were embedded within social and religious life at the time. While their significance is not questioned, the formal ceremonies that must have taken place at them are open to debate. On ritual sites, it is likely that the monuments were aligned with significant solar and lunar events. It is in the debate over such alignments and details of interior structures of a site in which a magnetometer survey can often provide important clues. These sites were often only sporadically used during prehistory, and therefore often produce very clean data sets. Under these circumstances, noninvasive surveying becomes very important.

An example of the detail that a magnetometer survey can contribute can be seen at Stanton Drew in southwest England. The site had been visited and described by many antiquarians, such as Aubrey and Stuckley, but had never been excavated. Visible today are three stone circles and associated avenues, the largest being the Great Circle, a set of monoliths aligned roughly circular. Some of the stones have fallen, and some appear to be missing. The geophysics section of English Heritage initially undertook a fluxgate gradiometer survey over the Great Circle (David et al. 2004). As the level of response was expected to be low, the surveyors adapted Geoscan Research FM instruments so that they could be carried as low as possible to the ground, and data were collected at 0.25 m along traverses separated by 1 m. The most obvious observation from this data set was that the stones were found to be encircled by a previously unknown ditch. Additionally, and much to the surprise of the surveyors, nine concentric rings of pits were found within the stone circle. In an effort to clarify the number and extent of the apparently concentric pit circles, an additional magnetometer survey was undertaken using a more sensitive cesium-vapor instrument (Scintrex Smartmag), and with a greater sample intensity

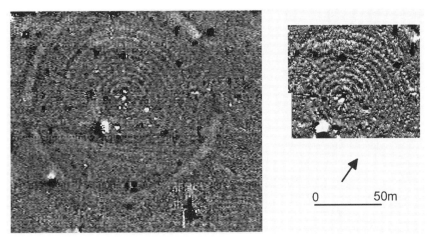

Figure 6.9. Magnetometer survey at the Great Circle at Stanton Drew prehistoric site, Somerset. *Left,* **data collected with a Geoscan fluxgate gradiometer, gray-scale −4 nT white to +4 nT black.** *Right,* **details within the circle revealed by use of a Scintrex Smartmag cesium-vapor magnetometer, gray-scale −2 nT white to +2 nT black. Images courtesy of English Heritage.**

of 0.125 m × 0.5 m (fig. 6.9). While individual anomalies in the outer ring could be easily resolved, the features with smaller diameter were not as easy to define (David et al. 2004). To overcome this problem, an automated routine was devised to determine objectively both the radii, and hence central positions, of all circles. In order to do this, the data set was processed with ZMT, and extreme values (>20 nT) were clipped. A reference anomaly for a pit was iteratively compared with each survey point. Once each pit location was identified, a modified Hough transform was used to fit circles to the data (Linford 2005). As a result of this analysis, the internal arrangement of this important site is now understood with some certainty.

While it is assumed that the individual pits are likely to have contained timber posts, there is still some debate as to the mechanism for the production of the magnetic contrast. It is possible that at least some of the timbers will have been burnt, although microbial action may be a more general pathway, possibly through magnetotactic bacteria in the postholes. What is certain is that high-sensitivity magnetometer surveying at similar sites is likely to provide startling images that help interpret ritual practice in prehistory. Using cesium-vapor magnetometers with a cart system over loess soils of very low intrinsic magnetic variation (i.e., low soil noise), Wolfgang Neubauer was able to unravel the internal structure of neolithic circular enclosures down to the layout of palisade trenches (Neubauer & Eder-Hinterleitner 1997). By comparing results from many sites in Austria, he was able to compile an archaeological assessment of this site type (Neubauer 2001b), including its orientation toward important stellar constellations (Kastowski et al. 2005). The detailed study also revealed severe erosion from agriculture on some sites, thereby assisting with Cultural Resource Management (Neubauer 2001a).

Historic Religious Sites

By way of contrast, historic religious centers are often well documented, and their use is similarly understood. In general, these sites have not been fruitful areas of endeavor for magnetic surveys, partly because of the construction materials used. In large parts of Europe, churches and monasteries have long been built in stone, and the preferred geophysical technique for mapping their remains is the earth-resistance survey. In other parts of the world, while the building techniques may be similar, no resistivity contrast may be encountered between the archaeological targets and the surrounding medium. In these cases, it is possible that magnetometer surveying may be a profitable investigation tool.

At Tell el-Balamun, Egypt, a British and Polish team have been examining a temple enclosure that dates from the twenty-sixth and the thirtieth dynasties. The site is defined by the temple enclosure wall, which measures 420 m × 450 m, and is too large for total excavation. In fact, after 12 years of fieldwork, the interior of the enclosure remained largely unknown (Herbich and Spencer 2006). It was expected that the buried remains would include mud-brick and monumental architecture fashioned out of stone. These structural elements are embedded in sand and the prospect of anyone undertaking an archaeologically productive earth-resistance survey at the site is low. However, there is considerable published literature that has shown the value of magnetometer surveying in the Nile Delta (Herbich 2003). In many cases, weak isothermally produced remanent magnetization is the cause of contrasts between mud-brick (Games 1977) and surrounding mud or sand.

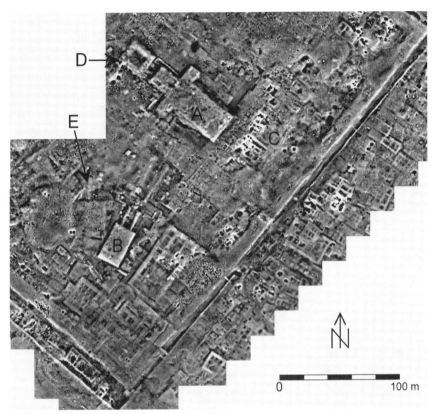

Figure 6.10.   An extract from a larger magnetometer survey at Tell el-Balamun, Egypt, gray-scale −15 nT white to 15 nT black. The survey shows known archaeological features within enclosure walls in detail: (A) temple of Nectanebo I and (B) Temple of Psamtik I, and many previously unsuspected features, (C) industrial zone, (D) potential barque station, and (E) another temple. Image courtesy of Tomasz Herbich.

Magnetic prospecting at Tell el-Balamun utilized Geoscan Research FM36 fluxgate gradiometers. An area in excess of 30 acres (13 ha) has been covered, with samples collected at 0.25 m intervals along traverses separated by 0.5 m, and in order to ensure high location precision, the individual 20 m × 10 m data grids were surveyed in parallel mode. It is clear from an extract of the total data that the magnetometer survey has mapped an enormous amount of detail from within the largely "blank" interior of the temple enclosure (fig. 6.10). It was to be expected that sacred and official buildings were to be found at this site, and evidently that was the case. The historically attested Temples of Nectanebo I (ca. 370 BC) and Psamtik I (ca. 650 BC) have been identified, but the level of survey detail surrounding them has provided new insights. At the rear of the Temple of Nectanebo I are strong responses pointing to an industrial zone, while at the front is a broad, square anomaly that indicates a low magnetic susceptibility contrast. It was anticipated that the latter was the result of a building constructed of limestone or sandstone. Subsequent excavation revealed that the structure was built of large limestone blocks in a foundation trench of mud-bricks and sand, and has been interpreted as a potential barque-station for the Temple of Amun, which lies to the west and is unsurveyed (Herbich and Spencer 2006, p. 19). For scholars interested in this site, the magnetometer survey has provided an overview that includes the mapping of known elements, as well as identifying completely new zones of activity, buildings, and even a completely new and unknown temple to the west of the Temple of Psamtik I. With this level of information, it is possible to redefine the research objectives at the site and at the same time be confident that excavations can be precisely placed.

Death and Burial

Although death is one certainty in life, burial in a fashion that will produce a response that can be measured using a magnetometer, or any other geophysical sensor, cannot be confidently predicted. The main reasons for this are that the body is largely an organic mass that contains very little magnetic material, and that, apart from the body, the grave is usually backfilled with the same material that was extracted during excavation. There is very little prospect of a significant magnetic contrast. However, metal grave goods can be good indicators, and this is one significant attribute that is commonly used to identify modern graves in forensic work (Cheetham 2005).

Occasionally a repeated pattern in the observed magnetic field is reported that can be attributed to graves cut directly in subsoil, but such situations are rare. In some parts of the world, or at specific periods, tombs have been built of stone or mud-brick, and in these cases magnetic contrasts can be identified. At the Northern Cemetery at Abydos, Egypt, excavation had barely investigated 2% of the presumed total area. In 2001, it was decided to evaluate the area by geophysical means (Herbich et al. 2003). As mud-brick was believed to be used in the construction of tombs in the area, it was decided to use an FM18 fluxgate gradiometer, with measurements taken on a 0.5 m grid. The survey was adjacent to a modern village, and to ensure the best data the area was cleared of modern debris prior to the survey. An extract from the resulting magnetometer survey reveals a wealth of detail, with shaft tombs scattered across the area (fig. 6.11).

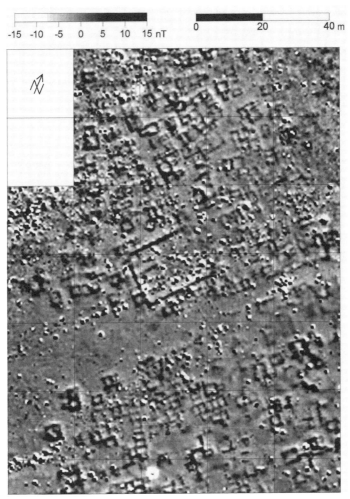

Figure 6.11. Fluxgate gradiometer survey of the North Cemetery, Abydos, Egypt (0.5 m data spacing interpolated to 0.25 m). The mud-brick construction of the tombs creates a significant magnetic contrast with the surrounding sandy matrix. The plethora of shaft graves indicate burials of lower-status individuals often overlooked amid the high-status monuments that dominate the historical record in this area. Image courtesy of Tomasz Herbich.

Key Points

Detailed magnetometer surveys may provide the wealth of data and examples that are needed to arrive at overarching archaeological interpretations for ritual sites, thus progressing far beyond the analysis of single sites. Most graves are refilled with the soil removed in their excavation, and hence usually show no notable magnetic contrast. Only where several weak anomalies appear to have been systematically aligned (as in a graveyard), or where graves were enhanced through stone, bricks, or ferrous grave goods is their detection in a magnetometer survey feasible.

Gardens

In his comprehensive handbook, *Garden Archaeology* (2005), the late Chris Currie provides a detailed description of the development of systematic garden archaeological methodology, from its origins in North America in the early 1960s to the recent use of multidisciplinary techniques in Britain by the author and his colleagues. Among the approaches he examined are geophysical prospection using earth-resistance, magnetometers, and induced-polarization surveys, and the more debatable method of dowsing. Currie's conclusion on these methods is that "geophysics should be used with caution on most garden sites. Both on sandy and clay soils, resistivity has been found to be wanting. It is not sufficient to claim success from the results of work where the only significant results are structural." It would appear that this view has been endorsed by Cole et al. (1997, 38), who wrote that geophysics is "vulnerable on sites where there is superimposition of features from later phases" and that "long-term potential of geophysical prospection to archaeology is still difficult to assess."

Undoubtedly gardens, in view of their generally transitory existence, present particular problems for archaeologists. While not unique, gardens encapsulate many aspects of shallow archaeological prospecting. As with so many archaeological sites, gardens are rarely finalized and fossilized after they have been made, and the original designs of gardens give way to those of later periods. In Europe, they have ranged from total formality with permanent structures and ornaments, through the natural landscaping of the eighteenth century, to the current trend toward formality, so that earlier features continue to be either destroyed or modified and integrated into later designs. In formal-garden planning, the frequent use of ironwork creates a framework for garden beds, themselves often taking the form of hedge-bordered parterres with graveled pathways. At the other extreme, landscaping gives rise to limited structures with the use of tree-lined avenues, natural-looking water features, and meandering, usually dry, pathways, and focal pleasure features. Clearly, any attempt at restoration of early gardens must define at what stage of development the restoration is to be focused, yet must record as much of the total garden history as possible. Currie's book demonstrates an acute awareness of these matters, but evidently over-anticipates the success of geophysical surveys relative to contributions to be made by other methods, such as botanical study or aerial photography, in the historical record of gardens.

In the following examples, the location and interpretation of two former formal gardens has been attempted using magnetometer and earth-resistance surveys. In the first case, Castle Bromwich Hall, the surveys were carried out over existing garden designs and in the second, Duncombe Park, an open lawned area was studied. Appropriate data processing was used to compare the two methods as to their relative effectiveness, and the correlation and discrepancies between them was examined in order to comment on the nature and history of the sites

Best Garden, Castle Bromwich Hall, Warwickshire

The geophysical survey at the hall was undertaken in 1989 by post-graduate students of the University of Bradford, West Yorkshire, as part of a major assessment of the gardens by Currie and his colleagues (Currie et al. 1991). Indeed, the outcome of the survey formed the basis of Currie's comments on the limitations of the geophysical methods used. The results were subsequently reappraised by Aspinall and Pocock (1995), making use of improvements in data processing procedures available by that time. It is of interest now to take one of the areas of survey, Best Garden, as an example, as it reveals typical problems associated with assessing a formal garden that developed over a period of two centuries.

At the time of survey, Best Garden was a square with sides of 50 m, directly facing the west front of Castle Bromwich Hall. It comprised four rose-beds set within a lawn and symmetrically disposed about a fountain. Each bed was approximately 18 m across, with its corners scooped out so as to give the appearance of a Maltese cross. For purposes of the survey, the full square area was divided into eight data grids, each wholly or partly of 20 m sides aligned with the square sides. Earth-resistance was measured using a twin-electrode system, with 0.5 m mobile probe-spacing, connected to a Geoscan Research RM4 meter with DL10 data logger. A Geoscan Research FM18 fluxgate gradiometer, hand-triggered at station markers, was used for the magnetic survey. In each case, data were recorded at 1 m intervals so that 400 readings were taken in each full data grid. It is clear that for both instruments the data ranges were unusually large. Earth-resistance values ranged over some 600 ohms, and magnetometer readings occasionally reached levels that saturated the instrument, resulting in artificial dummy readings appearing in the subsequent processing procedures. These magnetic readings undoubtedly came from extreme ferrous anomalies arising from existing iron garden features, and also from now-buried iron objects.

The data were reprocessed with Geoplot 3, which incorporates analytical and display facilities not available during the earlier interpretations. The results of the earth-resistance survey are given first so as to clarify the rose-bed layout. Figure 6.12a is a gray-scale image of the data after interpolation to quadruple the plotted data points. The wide earth-resistance range is such that only the lawn surrounding the beds is graded, suggesting that data within the beds require separate treatment. This is shown in figure 6.12b, where the earth-resistance data have been converted to conductance—a more realistic parameter under these extreme conditions. It can now be seen that the garden beds were surrounded, under the 1989 lawn, by high resistance, probably old graveled paths, and divided into the four sectors by north-south and east-west walkways. The Maltese-cross forms of the beds were at some stage simply squares, with possibly discrete "dry" edges. A circular area surrounding the present fountain formed a focal point of the garden, as confirmed below, in the magnetic survey. Each bed was patterned in a similar fashion, with discrete high-conductance scoops for planting arranged in a cross design surrounding a central scoop. The scoops evidently resulted from excavations refilled with more water-retentive organic soil than the surroundings.

The magnetometer results are presented in figure 6.12c and 6.12d, the former being raw data, with a wide gray-scale range, the latter after interpolation and over a more restricted range. Extremely high, discrete, ferrous anomalies dominate figure 6.12c to the extent that the characteristic neg-

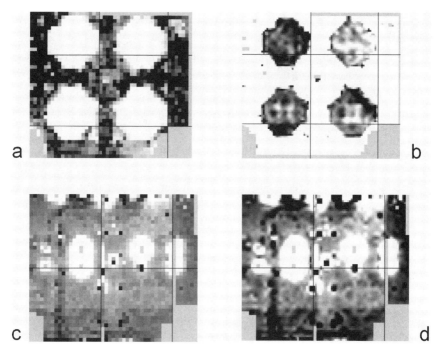

Figure 6.12. **Integrated geophysical survey of Best Garden, Castle Bromwich Hall, Warwickshire,** *north to left.* The twin-probe earth-resistance image (a) wide data range (100 to 800 ohm) shows outlines of flower beds in lawn settings. The earth-conductance image (b) over a restricted data range (90 to 150 × 10⁻⁴ mho) reveals details of beds. The magnetic image (c) over a wide data range (gray-scale −2 nT white to +30 nT black), shows anomalies probably from ferrous garden features. The magnetic image (d) with a restricted data range (gray-scale 0 nT white to +10 nT black) confirms the detail of structure seen in the conductivity image.

ative haloes effectively obliterate adjacent weak anomalies over a large part of the eastern section of the survey. The replacement of saturation values by dummy readings, particularly on the north-south pathway axis, should be noted. Furthermore, the coarse sampling interval of 1 m has led, on some occasions, to the nonmeasurement of the central positive peak although the negative haloes are clearly present. It can be seen that the radial paths are edged at regular intervals by pairs of discrete ferrous anomalies, suggesting the presence, in an earlier garden design, of iron-footed arches or pergolas spanning the paths to carry climbing plants. The central fountain also has associated ferrous anomalies, presumably from iron service pipes. The eastern edge of the garden is dominated by large ferrous anomalies associated with iron features at the adjacent west front of the hall. By interpolating the raw data and contracting the gray-scale range (fig. 6.12d), a significant pattern is revealed in the southwest bed that coincides well with that of the conductance data. The scoops are seen here as positive magnetic anomalies consistent with an organic soil fill, and there is some evidence in this bed, and that in the northwest, of the presence of the corner features seen in figure 6.12a. Two parallel, linear, magnetic anomalies run east-west through the two northern data grids. The weaker, outer line appears to turn south at its ends, suggesting a drainage channel associated with the peripheral paths. The inner, stronger anomaly may be linked with the relics of an excavated, substantial brick wall, at varying depths up to one meter, identified as an eighteenth- to nineteenth-century feature. Deeper excavated features were not, in the main, detected, although a significant, broad band of low-resistance and enhanced magnetization passing north-south at the western extreme of the survey is of some interest. A sample trench dug in this area revealed the presence of a quarry hollow 5 m wide and 1 m below the present surface, and was identified as a seventeenth-century feature.

In the light of this reappraisal, it is evident that Currie's frustration at the Castle Bromwich site was due in part to the lack of high-quality graphical presentations that have since become routine. However, any attempt to establish a detailed time sequence for developments of the garden layout, in the absence of documentation and excavation, must be purely conjectural.

Croquet Lawn, Duncombe Park, North Yorkshire

The early eighteenth-century mansion of Duncombe Park has, in front of its eastern façade, a large rectangular lawn extending some 60 m north-south across the house frontage and stretching eastward approximately 80 m. A feature at the eastern end is an eighteenth-century sundial, centrally placed, so that a line from the center of the portico of the house to the sundial bisects the lawn. The lawn itself is dished (i.e., sunk) to a depth of 0.75 m, and currently used for croquet.

Aerial photographs revealed faint traces of linear features within the lawned area, and a geophysical survey was undertaken in 2005 in an attempt to reveal the nature of these marks (Wheeler et al. 2007). Although there has been periodic restructuring of the house since its construction,

early prints of the area of the east frontage revealed no evidence of cultivation. However, it is recorded that the building was used as a girls' school
from the end of World War I to the mid-1980s, when it was reoccupied
by the sixth Lord Feversham. To facilitate the survey, the lawn was divided
into Data Grid squares with sides of 30 m to complete the coverage of the
area. Magnetometer and earth-resistance surveys were used, and the opportunity was taken to assess the mobile sensor platform (MSP), produced
by Geoscan Research (Walker et al. 2005), which had been upgraded to
include an FM256 fluxgate gradiometer mounted on the cart. Earth-
resistance was measured via the four spiked wheels of the cart at 0.75 m
separation to form a square array (Aspinall and Saunders 2005), the data
being logged through an RM15 earth-resistance meter. The platform was
pulled in a zigzag manner over the survey area, with a 1.0 m traverse spacing. Individual readings were recorded at 0.25 m with the magnetometer,
and 0.5 m with the earth-resistance instrument.

As for Castle Bromwich, all data were processed using the Geoplot 3
program and presented as gray-scales. To facilitate easy comparison of the
two surveys, the earth-resistance readings were converted to conductance
values before plotting. These were interpolated once in the Y-direction to
produce a square data matrix of 0.5 m, and are displayed in figure 6.13a.
It can be seen that high-conductance linear anomalies create two rectangular enclosures approximately 50 m long and 15 m wide. There is some
ambiguity at the eastern end of the rectangles, where in both cases further
north-south terminals occur to the east of the main areas. Parallel, high-
conductance lines run northwest-southeast, in an alignment typical of wet
drainage gullies, confusing the rectangular pattern. There is also some evidence in the rectangles of discrete areas of high conductance.

The raw magnetometer data are imaged in figure 6.13b, and it can be
seen that the zigzag procedure has resulted in marked striping of the responses along the lines of traverse. As discussed in chapter 4, this usually
arises from an imbalance in the setup of the instrument. In this case, however, it was decided that the effect was probably from siting the magnetometer on the cart in the vicinity of a small ferrous component. The
normal option for striping removal, with Geoplot, is that of zero-mean traversing (ZMT), containing a data window of $\pm Z$ nT, within which to establish the ZMT correction. Evidently the rejection of minor linear features
along the traverse depends on the window Z chosen. Thus figure 6.13c

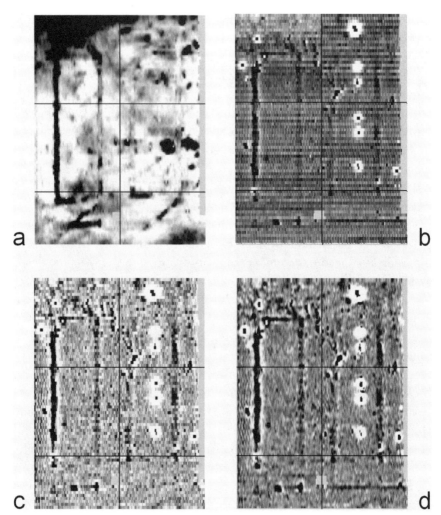

Figure 6.13.   Integrated geophysical survey of the croquet lawn, Duncombe Park, North Yorkshire, *north to right*, all gray-scale images. (a) Earth conductance data derived from mobile square array resistance survey: data range 25 to 115 ($\times$ 10^{-3}) mho. (b) Magnetometer survey; raw data showing significant striping: data range +5 nT white to +5nT black. (c) Magnetometer data after ZMT, with a window of ±5 nT: gray-scale data range −1 nT white to +5nT black. (d) Magnetometer data after ZMT, with a ±2 nT window: gray-scale data range −2 nT white to +5 nT black. Note the retention of a north-south feature prominent in the raw data.

is produced with a window of ±5 nT and the stripes are virtually eliminated. However a prominent linear anomaly, seen in the raw data running north to south in the northeast data grid, is lost. There is little doubt that the anomaly stands above the background and has been lost in processing. However, if the window is reduced to ±2 nT (fig. 6.13d), this anomaly is retained and matches a similar anomaly present in the data grid opposite. Thus the linear edges of the two rectangular beds appear as high-conductance, positive magnetic anomalies typical of trenches cut into subsoil and bedrock and refilled with humus-rich material for plant culture. The western ends of the rectangles are more diffuse, however, suggesting surface intrusion. It is interesting to note that the two "inner" eastern extremes seen in the conductance image do not appear in figure 6.13d, although there is faint magnetic evidence of extension of the beds to the eastern limits.

It may also be significant that a broad band, which partially obliterates the east-west anomalies in both surveys, crosses the two eastern extremes: this may represent later site disturbance. A line of prominent, ferrous, bipole anomalies runs east to west through the northern bed in the form of three pairs, the anomalies in each pair being separated by 13 m. This is close to the recognized separation of the supporting posts of tennis nets. In view of the use of the house as a girls' school in the twentieth century, it is reasonable to conclude that three tennis courts were laid out in the northern half of the lawn.

The investigation suggests that, probably sometime in the eighteenth century, the eastern lawn at Duncombe Park was laid out as two formal rectangular beds symmetrically disposed on either side of the line from house to sundial, the beds being probably edged by shrubs requiring a depth of planting. Later, in the fashion of the period, the lawn was created for recreational purposes, culminating in its present use as a croquet lawn.

Key Points

This study of garden features has provided a set of circumstances that illustrate specific difficulties in the interpretation of magnetic data. The surveys often only cover a small area, surrounded by current garden furniture, and the targets may be slight and confused by later remodeling. As a result, analysis is often difficult but ultimately rewarding. Some key points arising from these studies are:

Relevant anomalies range from weak to very strong in magnitude, and this must be reflected in the technique and methodology employed. Gradiometers with shorter base lengths are preferable to suppress the influence of nearby ferrous objects.

Surveys are frequently undertaken over small areas encumbered by upstanding obstacles, requiring a low-technology approach, such as hand data-logging and highly portable magnetometers.

Interpretation of magnetic data is often difficult without the aid of other geophysical techniques, such as earth-resistance, or indeed supporting historical evidence. A full appreciation of the geophysical data is very difficult without an adequate knowledge of a garden's history and, ideally, its archaeology.

Igneous and Ferrous Anomalies

Rocks are classified as igneous, sedimentary, or metamorphic on the basis of their origins. Most igneous rocks have crystallized from a high temperature, molten silicate liquid or magma. They can be classified as intrusive if they have solidified beneath the earth's surface, typically as granites, or extrusive when they have solidified after emerging from a volcano or other vent, typically the basalts. Ferromagnetism is present in both forms, the basalts usually being the more magnetic. As we have seen in chapter 1, the cooling of such rocks through their Curie temperatures results in remanent magnetization in the direction of the earth's magnetic field at the time. Over the course of geological time, igneous formations may outcrop at, or near to, the earth's surface in the form of near-horizontal layers called sills, or vertical slabs termed dikes. On a larger scale, they may be present as volcanic plugs intruding upward through the earth's surface. Igneous rocks are readily detected in all these forms by magnetometers through their strong magnetization. The same applies to glacial erratics created as a result of the breakup and dispersal of glacial rocks, from their places of origin, by the action of ice flows.

Other magnetic anomalies have their origin in metallic iron in its various forms. They are characterized by very high magnetic susceptibility values, leading to readily induced magnetization in the ambient earth's magnetic field. In some of its forms, such as hard steel, permanent magnetism may additionally be present. The degree of magnetization in an iron object is highly directional, depending on its shape and orientation in a magnetizing field. Thus a thin rod of iron becomes highly magnetized in a field along its axis while a plate in a field perpendicular to its plane exhibits little magnetization. As we have seen, it is often the presence of iron artifacts in the soil that is responsible for the masking of sought-after features on an archaeological site. Igneous features can also greatly affect the acquisition of relevant archaeological data. In the site examples that follow, the influences of igneous and ferrous anomalies are illustrated—not always to the detriment of feature interpretation.

Rothiemay Recumbent Stone Circle, Banffshire

A recumbent stone circle (RSC) comprises a ring of substantial stones graded upward in height from north to south, terminating in a large recumbent stone (the "altar"), which lies at the southeast to southwest of the monument and is flanked by the two tallest stones. Typically, twelve such stones form a ring roughly circular and up to 35 m in diameter. Frequently, an inner ring of smaller stones forms a central cairn. More than seventy of these impressive monuments are sited to the south of the Moray Firth and east of the foothills of the Grampian Mountains of Scotland. Many of the circles have been slighted, or damaged, either for stone use, land clearance, or religious motives, so that few intact examples remain. The Rothiemay RSC is located at the southern tip of the Portsoy (Moray Firth)-Rothiemay tongue of igneous intrusive geology east of the village of Milltown of Rothiemay. Only four upright stones remain, two on either side of a massive recumbent stone that has no flankers and lies at the south southwest extreme of the rough arc formed by the uprights. In 1998, a geophysical survey was carried out (Aspinall 2006) using earth-resistance and magnetometer techniques in an attempt to locate the siting of the lost stones of the monument. Burl (1976) estimated that the circle diameter had been 28 m. An earlier researcher (Cole 1903) proposed that the recumbent stone lay within the ring and in line with a ring of smaller stone blocks laid on edge.

The significant revelation in the 1998 survey, that all the standing stones are strongly magnetic, and of obvious igneous origin, enabled the interpretation of other discrete, high-value magnetic anomalies as the relics of other upright stones now buried or broken below ground level. A version of the image of magnetic anomalies is shown in plate 7. Here the unit Data Grid squares are of 10 m a side, and the Geoscan Research FM 36 fluxgate gradiometer readings have been taken at 0.5 m intervals. The original data set was de-spiked to remove any single-valued anomalies, and then an appropriate palette was used to display only extreme values of the positive (red) and negative (blue) anomalies. Dummy blocks, from the presence of the existing stones, can be seen surrounded by strong anomalies of a complex nature since the tops of the stones exceeded the gradiometer's upper sensor in height. It can be seen that extended but finite magnetic anomalies are oriented quite randomly with respect to magnetic north. Some are inverted, while others suggest stones inclined at different angles to the horizontal. It is clear, however that these orientations refer to the magnetic axes of the stones, rather than the geometric. Although the site was extremely wet, with pools of water surrounding the standing stones, the presence of high-resistance responses at the sites of some magnetic anomalies gave strength to the suggestion of buried stones features. As a result of the survey, it was proposed that the monument probably comprised a roughly circular ring of stones, 33 m in diameter, enclosing the recumbent stone. This formed the focus of an inner stone horseshoe some 20 m across, with its open end to the north and surrounding a magnetic (and high-resistance) platform—possibly a cairn. Such a configuration compares well with other recumbent stone circles of the area.

Towton Battlefield, North Yorkshire

The battle of Towton was fought in 1461 during the English War of the Roses and was centered on a field between the villages of Towton and Saxton. The contestants were the Lancastrian and Yorkist armies, and the battle is recorded as the bloodiest fought on British soil. The two armies were drawn up on either side of a shallow east-west valley, Towton Vale, with the Yorkists, the final victors, on the southern slopes. A significant factor in this, and other fifteenth-century warfare, was the use of archers by both sides, and it is reasonable to assume that thousands of arrows would have been fired at the outset of the battle, before the inevitable close-hand fighting.

The residue of the battle has long been the focus of activities of metal-detector-using artifact hunters, and in recent years some have adopted a very responsible approach to their investigations, providing valuable data for the reconstruction of the battle (Sutherland and Schmidt 2003; Sutherland 2007). Consequently, finds of arrowheads have been systematically recorded in terms of numbers and locations, confirming the concentrations to be expected from salvoes of arrows. A magnetometer survey was carried with a Geoscan Research FM256 twin-gradiometer system over the central area of the battlefield. Data readings were taken at 0.25 m intervals and a 1 m traverse separation. The main purpose was to locate the presence of mass battle graves reputedly dug at the site. Plate 8 is a gray-scale image of the final data, corrected for slight difference in the two gradiometer performances using the ZMT processing tool of Geoplot 3. The surface geology at the site is of magnesian limestone overlaid by thin topsoil. The image is unrewarding from the point of view of sought-for archaeological features. It is dominated by strong cultivation patterns that are more prominent to the south in the valley bottom, where there is a greater deposition of high magnetic susceptibility soil. However, in view of the earlier records of arrowhead finds, the magnetic data were further processed to isolate spikes generated by iron. To that end, the data set in Plate 8 was subjected to a low-pass filter using four outliers on each side of a data point in the traverse (high sampling density) direction and one perpendicular to the traverse. This set was subtracted from the initial data set so that only the highly discrete anomalies remained. These were selectively examined, using the display palette employed on the Rothiemay data, with limits of ±3 standard deviations around the mean of the treated data set. The residual anomalies are both positive and negative, corresponding to a random assembly of small ferrous fragments, a proportion of which might be expected to be arrowheads. When compared with the earlier collection of arrowheads from the site, it is clear that the sparsely populated parts of the survey correspond very well with the concentration of the collection from the site. This gives confidence to the procedure as a means of investigating early battlefields. Obviously, the anomalies recorded will include other small ferrous objects, such as agricultural debris, but this "waste" will also form part of a metal-detector surveyor's collection. Indeed, in a different archaeological context, such finds may have a cultural significance.

Key Points

Igneous and ferrous features produce very distinct magnetic anomalies in terms of signal strength and the shape of recorded readings. Data processing therefore has to be specifically tailored to these anomalies to identify the causative features.

The strong magnetization of such features can be an advantage, often resulting in considerable magnetic contrast and hence improving their detectability.

CHAPTER SEVEN

MAGNETOMETRY AND ARCHAEOLOGY: THE FUTURE

The thrust of this book has been to establish a framework for the significance and understanding of magnetometry with respect to archaeological investigations. While we have detailed the theory and methods of magnetic techniques as they are employed at present (2007), it is certain that important changes will take place in the near future. Currently, we are in a particularly productive period of research and development in magnetic surveying that is reflected in the pages of the field's flagship journal, *Archaeological Prospection*, as well as in the renewed vigor of survey-instrument manufacturers. Recent developments, although not fully formalized, allow us to indicate where some of the more substantive changes may occur.

Sensors

One key change during the last two or three years has been the development of different sensors and their greater availability. In the relatively recent past it was common to find fluxgate or cesium vapor magnetometers (CV*s*) unavailable for rent or purchase, except in particular countries. Now, both types of instruments are offered locally for archaeological and other work by dealers in many places across the globe. This has allowed surveyors to define the scope of surveys more precisely, i.e., they can match an available sensor to the questions that are asked, rather than make do with what they have. The archaeological benefit of this change is very obvious, and bodes well for the integration of magnetometer surveys into complex and forward-looking research strategies, including the development of new magnetometer systems themselves.

The compact size and vector-measuring property of a fluxgate sensor lends itself well to the development of a triaxial assembly comprising three fluxgates mounted mutually perpendicular to each other. Its output vectors give a total-field (intensity) measurement, i.e., the fluxgate measurement then becomes equivalent to that of a CV. Systems such as the Bartington Mag-01 series instruments (www.Bartington.com), which at best have a range of ±70μT, have been in use for a number of years, with applications in aspects of geophysical exploration. A three-component fluxgate magnetometer developed more recently in Japan has been used in archaeology, originally in the investigation of a burial mound (Nishimura et al. 1996). In that case, however, the three (x, y, z) components of the earth's magnetic flux density were measured separately to provide analytical information on the three specific anomalies. The construction of a triaxial fluxgate magnetometer for continuous intensity measurement at the sub-nanotesla level presents a number of design difficulties. They are not insurmountable, however, and would lead to a highly versatile and portable system.

Similar innovations may also appear based on the magnetoresistive properties of some electrically conducting compounds. This phenomenon has been known for many years in terms of the increase in electrical resistance of the element bismuth in the presence of an external magnetic field of large magnitude. The development of conductive compounds with much higher sensitivities to external fields has led to investigations of a magnetoresistive vector magnetometer operating in the micro to nanotesla range (Smith et al. 1991), and commercial sensors are currently available with this sensitivity. Once again, however, application to archaeological prospection awaits the design of systems with appropriate sensitivity that are rugged and stable in the field.

The most promising development in the past few years, perhaps, has been that of a SQUID-based vector gradiometer system (see chapter 2) that has been successfully field-tested. The increased sensitivity that can be obtained using SQUID technology suggests that archaeology that was invisible to other magnetometers may now be recognizable. Again, the analytical value of data from very sensitive magnetometers may reveal detail that allows a more complete vision of what now lies buried in the subsurface.

Multiple Sensors, Carts, and Sledges

Until the late 1980s, it was conventional practice for single magnetometers or gradiometers to be handheld, with stationary measurements collected over a relatively wide mesh. During the last decade or so, however, we have seen the development of multiple sensors on handheld instruments, and this has significantly improved the area that can be covered during a working day. The 1.0 m separation Bartington fluxgate instrument, described in chapter 2, is frequently used as a dual-measurement device, and is often employed for commercial surveys in that version (fig. 7.1). The lightweight nature of such systems ensures that even relatively rugged terrain can be assessed for archaeological remains.

Figure 7.1. The Bartington Grad601-2 handheld (1.0 m) fluxgate system collects measurements on adjacent traverses separated by 1.0 m, although smaller inter-transect distances can be achieved.

Figure 7.2. The Geoscan Research MSP40 Mobile Sensor Platform integrates both resistance and fluxgate (0.5 m vertical separation) measurements.

Manufactures have led the way, however, by producing wheeled and sledge-mounted carrying platforms (fig. 7.2). For CV systems in particular, the trend is for increasing numbers of sensors and greater data density (fig. 7.3). The technical puzzles related to these devices can be complex, but it is obvious that many research groups across the world have found little difficulty in solving them, although the devices still cannot ordinarily be used on rugged terrain.

Figure 7.3. The Geophysical Exploration Equipment Platform (GEEP) system, developed by the University of Leicester and Geomatrix Earth Science Ltd., connects many types of sensors (CV array shown here) for towing on a machine-pulled sledge.

Measurement Location

A major factor in methodological advances in the development of procedures for the collection of data has been the incorporation, in geophysical systems, of real-time, high-precision GPS instruments (fig. 7.4) into surveys. This is particularly important as this allows both the direction of survey and three-dimensional location of measurements to be recorded. A consequence is that grid-free data collection strategies can be implemented, reducing the time of fieldwork, as well as allowing instruments to be wheeled or dragged across a landscape at high and variable speeds. This is a welcome aspect of magnetometer surveying, and it is likely that GPS will become an established component of geophysical fieldwork. An added bonus is that topographic information

Figure 7.4. This cart-carried, three-channel fluxgate system manufactured by Foerster Instruments Inc. uses real-time GPS for highly accurate measurement location.

can be collected at the same time, allowing a traditional aspect of archaeological analysis to inform the surveyor directly of the significance of the data (figure 7.5).

Analysis and Visualization of Large Data Sets

Given that surveys of increasing data density are likely to be conducted in the future, and that multiple data types can be collected in a single sweep of an area, a question that needs to be considered is, how are they going to be visualized? Throughout the brief history of magnetometer surveying there has been ever-increasing sophistication of graphics—often to good effect, when we consider the re-analysis of the garden case study in the previous chapter—but how can we get overviews of large

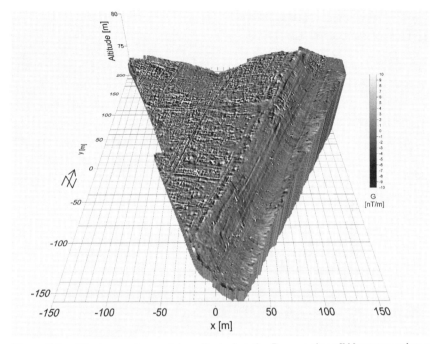

Figure 7.5. **The image shows data collected at the Roman city of Wroxeter using a SQUID system.** The magnetometer data were captured at up to 28 mph (45 km/h) using real time GPS and draped over topographic information collected at the same time. Image courtesy of Volkmar Schultze.

data sets and display detail that may be archaeologically important? GIS is now routinely used across archaeology, and allows integration of many aspects of the archaeological record. That is a welcome step in that the ability to analyze magnetic data within GIS environments suggests that the technique will remain embedded within archaeological research strategies. Of course, many workers produce excellent work using CAD, which links back to the engineering history of the subject, and the inherent plan-based nature of CAD output is crucial in many commercial applications. While visualization is a constantly changing delivery environment, the printed version remains problematic; the days of only printed delivery are limited, and it is likely that more versatile, open access to data will be part of the solution. In some countries, national archives are beginning to accept geophysical data, so the revisiting of legacy data sets will be possible.

Archaeological Questions: Landscapes and Low-Contrast

Are there any changes, from the point of view of the archaeologist, to the questions that we are likely to be posed in the future? Remote-sensing in general is increasingly required to provide data that will answer questions in the following archaeological scenarios.

Firstly, archaeologists are becoming increasingly interested in investigating landscapes. The work described in Powlesland et al. (2006), in which 2,500 acres (1,000 ha) have been surveyed using handheld devices, is a good example (see fig. 7.6). Given that immense achievement, how

Figure 7.6. A large-area fluxgate magnetometer survey image showing track-ways, land divisions, and burial monuments of late Iron Age and Roman settlement in the Vale of Pickering, North Yorkshire. Data gray-scale is −10 nT white to +10 nT black. Image courtesy of The Landscape Research Center.

many more acres or hectares could be collected using automated systems? Even with the current generation of multiple-sensor systems, it is likely that coverage could be doubled in the same time, and with a vehicle-driven, cart-based system the speed of collection could be many times that available now.

Secondly, there are some research areas where traditional magnetometry has provided little or no information of archaeological worth, yet they are areas where the new instruments that are becoming available could re-ignite interest. For example, there are sites that have proved difficult to map because of low magnetic contrasts between features and surrounding soil or the depth to the target. In this context, the development of the SQUID system has undoubtedly led to a new range of possibilities as a research tool in archaeo-prospection. A single-sensor SQUID is too sensitive for the rugged conditions of the field, both in terms of terrain and operator, but as a true gradiometer it opens a new vista for the interpretation of magnetometer data. We have seen in chapter 2 that the SQUID gradiometer is a vector instrument and will measure the gradient of the component of the earth's field in a specific direction.

Field tests carried out by a team at Jena, Germany, (Chawla et al. 2003) involved two approaches: a low-temperature system (LTS) to measure dB_x/dz, where B_x is the horizontal component of the earth's flux density, coinciding with the direction of traverse, and z is the vertical direction; and a high-temperature system (HTS) to effectively measure an approximate value of dB_z/dz over parallel traverses. The sensor separations were very different for the two tests, however: for the LTS, the base line was 4 cm, while for the HTS it was 60 cm, the former obviously giving a better measure of true gradient. Further, the resolution of the LTS was quoted as 2 pT/m while that of the HTS was 330 pT/m (1 pT = 10^{-3} nT). The difference between the two images of the test site, produced in the 2003 report, is striking, even accepting the practical problems encountered with the HTS. Whereas the image produced by the HTS is characteristic of traditional gradiometers, that of the LTS is characteristic of a directional (along the traverse line) high-pass filter, with the consequent appearance of high-definition shadows. It is clear that the interpretation of such data will require a greater appreciation of the nature of such magnetic anomalies than is currently recognized with gradient magnetometers.

The exceptional sensitivity of the SQUID leads us to the possibility of investigating low-contrast sites, those with very small magnetic susceptibility contrast at shallow depths, such as those of postholes, or the actual depth to archaeology. For a dipole source, we have seen that the magnitude of flux density falls off as the inverse cube of the depth to the source: a magnetic flux density of 1 nT at 1 m depth becomes 0.125 nT at 2 m. Evidently archaeological features will approximate the dipole model at greater depths. Thus the advent of higher sensitivity magnetometers should present us with a new area of study over deep but magnetically quiet sites, such as desert locations and also valley bottoms with high alluvial overlays.

Use of the SQUID "true" gradiometer would seem to be very attractive for deep site investigations. However, if the dipole field falls off as the inverse cube of the distance, that of its gradient will fall off as the inverse fourth power of depth. In the above example, this means the gradient will reduce by a factor of $\frac{1}{16}$. This would suggest that a CV magnetometer, operated in differential mode, would perhaps give a better performance for such sites than the current design of SQUID gradiometer. However, SQUIDs for archaeological prospection are at an early stage of development, and compensation by changes of sensor separation for anomalies at increasing depth is a clear way forward to revealing archaeology hitherto unlocatable by traditional sensors.

Some Final Thoughts

As archaeological geophysicists, we have traveled some distance from measuring magnetic fields with a compass. The latest magnetometers are measuring devices that are embedded in many archaeological projects, and the case studies that we have drawn on show the varying nature of the responses that can result from archaeological, geological, geographical, depth-related, and instrument-inherent factors. To understand and interpret the magnetic data that are now measured requires a knowledge that spans classical and quantum physics, and an understanding of archaeology in its broadest sense as well. The need for rapid, high-density, high-resolution magnetometer surveys will increase in the future as the search for diminishing archaeological resources grows ever more important.

REFERENCES

Aitken, M. J.
 1958 Magnetic prospecting: I—The Water Newton survey. *Archaeometry* 1:24–29.
 1986 Proton magnetometer prospection: Reminiscences of the first year. *Prospezioni Archeologiche* 10:15–18.
 1961 *Physics in Archaeology.* New York: Interscience.

Aitken, M. J., and J. C. Alldred
 1966 Prediction of magnetic anomalies by means of a simulator. *Prospezione Archeologiche* 1:67–71.

Alldred, J. C.
 1964 A fluxgate gradiometer for archaeological surveying. *Archaeometry.* 7:14–19.

Allum, G. T., R. G. Ayckroyd, and J. G. B. Haigh
 1995 A new statistical approach to reconstruction from area magnetometer data. *Archaeological Prospection* 2 (4):197–205.

Aspinall, A., and J. A. Pocock
 1995 Geophysical prospection in garden archaeology: An appraisal and critique based on case studies. *Archaeological Prospection* 2 (2):61–84.

Aspinall, A., and M. Saunders
 2005 Experiments with a square array. *Archaeological Prospection* 12 (2): 125–129.

REFERENCES

Aspinall, A.
2006   Magnetic stones: An investigation of the recumbent stone circle at Rothiemay, Banffshire. In *Going over old ground: Perspectives on geophysical and geochemical surveys in Scotland*, R. E. Jones, and L. Sharpe, eds. Oxford: BAR, Archaeopress.

Bartington, C., and C. E. Chapman
2004   A high-stability fluxgate magnetic gradiometer for shallow geophysical survey applications. *Archaeological Prospection* 11(1):19–34.

Becker, H.
1995   From nanotesla to picotesla: A new window for magnetic prospection. *Archaeological Prospection* 2 (4):217–228.
2001   Duo- and quadro-sensor configuration for high-speed/high-resolution magnetic prospecting with cesium magnetometer. In *Magnetic Prospecting in Archaeological Sites*, H. Becker and J. W. E. Fassbinder, eds. Munich: ICOMOS.

Becker, H., and J. W. E. Fassbinder
1999   Magnetometry of a Scythian settlement in Siberia near Cicah in the Baraba Steppe, 1999. In *Third International Conference on Archaeological Prospection*. Munich: Bayerischen Landesamtes für Denkmalpflege.

Bednorz, J. G., and K. A. Müller
1986   Possible high-T_c superconductivity in the Ba-La-Cu-O system. *Zietschrift für Physik B Condensed Matter* 64 (2):189–193.

Benech, C.
2007   A new approach to the study of city planning and domestic dwellings in the ancient Near East. *Archaeological Prospection* 14 (2):87–103.

Bescoby, D. J., G. C. Cawley, and P. N. Chroston
2004   Enhance interpretation of magnetic data using artificial neural networks: A case study from Butrint, Southern Albania. *Archaeological Prospection* 11 (4):189–200.

Bevan, B. W.
1994   The magnetic anomaly of a brick foundation. *Archaeological Prospection* 1 (2):93–104
1995   Geophysical Prospecting. *American Journal of Archaeology* 99:88–90.

Blakely, R. J.
1996   *Potential theory in gravity and magnetic applications.* Cambridge: Cambridge University Press.

Burl, A.
1976   *The stone circles of the British Isles 167–187.* London and New Haven: Yale University Press.

Chávez, R. E., A. Tejero, L. Barba, and L. Manzanilla
2001   Site characterization by geophysical methods in the archaeological zone of Teotihuacán, Mexico. *Journal of Archaeological Science* 28:1265–1276.

Cheetham, P.
2005   Forensic geophysical survey. In *Forensic archaeology: Advances in theory and practice.* J. R. Hunter and M. Cox, (eds. London: Routledge Press).

Chwala, A., R. Stolz, R. IJsselsteijn, V. Schultze, N. Oukhanski, H.-G. Meyer, and T. Schuler
2001   SQUID gradiometers for archaeometry. *Superconductor Science and Technology* 14:1111–1114.

Chwala, A., R. IJsselsteijn, T. May, N. Oukhanski, T. Schüler, V. Schultze, R. Stolz, and H.-G. Meyer
2003   Archaeometric prospection with high-T$_c$ SQUID magnetometers. *IEEE Transactions on Applied Superconductivity* 13:767–770.

Ciminale, M., and M. Loddo
2001   Aspects of magnetic data processing. *Archaeological Prospection* 8 (4): 239–246.

Clark, A.
1990, 1996   *Seeing Beneath the Soil.* London: Batsford Press.

Clark, A., and D. Haddon-Reece.
1972–73   An automatic recording system using a Plessey fluxgate gradiometer. *Prospezioni Archeologiche* 7-8:107–113.

Cole, F. R.
1903   The stone circles of northeast Scotland chiefly in Auchterless and Forgue. *Proceedings of the Society of Antiquaries of Scotland* 37:134–137.

Cole, M. A., A. E. U. David, N. T. Linford, P. K. Linford, and A. W. Payne
1997   Nondestructive techniques in English gardens: Geophysical prospecting. *Journal of Garden History* 17:26–39.

Crowther, J.
2003   Potential magnetic susceptibility and fractional conversion studies of archaeological soils and sediments. *Archaeometry* 45:685–701

REFERENCES

Crowther, J., and P. Barker
1995   Magnetic susceptibility: Distinguishing anthropogenic effects from the natural. *Archaeological Prospection* 2 (4):207–215.

Currie, C. K.
2005   *Garden archaeology: A handbook.* York: Council for British Archaeology.

Currie, C. K., and M. Locock
1991   An evaluation of a range of archaeological techniques used at the historic garden of Castle Bromwich Hall, 1989–1990. *Garden History* 19 (1):77–99.

David, A.,M. Cole, T. Horsley, N. Linford, P. Linford, and L. Martin
2004   A rival to Stonehenge? Geophysical survey at Stanton Drew, England. *Antiquity* 78 (300):341–358.

Desvignes, G., A. Tabbagh, and C. Benech
1999   The determination of the depth of magnetic anomaly sources. *Archaeological Prospection* 6 (2):85–106.

De Vore, L., T. N. Smekalova, and W. Bevan
1994   A comparison of magnetic gradiometers. Mount Hutton, Australia: Ultramag Geophysics Pty.

Dittmar, J. K., and J. E. Szymanski
1995   The stochastic inversion of magnetic and resistivity survey data using the simulated annealing algorithm. *Geophysical Prospecting* 43 (3): 397–416

Dittrich, G., and U. Koppelt
1997   Quantitative interpretation of magnetic data over settlement structures by inverse modeling. *Archaeological Prospection* 4 (4):165–178.

Eder-Hinterleitner, A., W. Neubauer, and P. Melichar
1996   Restoring magnetic anomalies. *Archaeological Prospection* 3 (4): 185–197.

Eder-Hinterleitner, A., and W. Neubauer
2001   Reconstructing neolithic ditches by magnetic modeling. In *Filtering, Optimization and Modeling of Geophysical Data in Archaeological Prospecting*, M. Cucarzi and P. Conti, eds. Milan and Rome: C. M. Lerici Foundation.

Fassbinder, J. W. E., H. Stanjek, and H. Vali
1990   Occurrence of magnetic bacteria in soil. *Nature* 343:161–163.

Fassbinder, J. W. E., and H. Stanjek
1993   Occurrence of bacterial magnetite in soils from archaeological sites. *Archaeologia Polona* 31:117–128.

Gaffney, C. F., and J. A. Gater
1993   Development of remote sensing, part 2: Practice and method in the applications of geophysical techniques in archaeology. In *Archaeological resource management in the UK,* J. R. Hunter and I. Ralston, eds. Stroud, Gloucestershire: Alan Sutton Publishing.

Gaffney, C. F., J. A. Gater, P. K. Linford, V. L. Gaffney, and R. White
2000   Large-scale systematic fluxgate gradiometry at the Roman city of Wroxeter. *Archaeological Prospection* 7 (2):81–89.

Gaffney, C. F., J. A. Gater, and S. M. Ovenden
2002   The use of geophysical techniques in archaeological evaluations. *Institute of Field Archaeologists* Paper No. 6.

Gaffney, C. F., and J. A. Gater
2003   *Revealing the buried past: Geophysics for Archaeologists.* Stroud, Gloucestershire: Tempus Publishing.

Games, K. P.
1977   The magnitude of the palaeomagnetic field: A new nonthermal, non-detrital method using sun-dried bricks. *Geophysical Journal of the Royal Astronomical Society* 48:315–330.

Graham, I. D. G., and I. Scollar
1976   Limitations on magnetic prospection in archaeology imposed by soil properties. *Archaeo-Physika* 6:1–124.

Haigh, J. G. B.
1992   Automatic grid-balancing in geophysical survey. In *Computer applications and quantitative methods in archaeology 1991,* G. Lock and J. Moffett, eds. Oxford: BAR.

Heathcote, C.
1983   Applications of magnetic and pulsed induction methods to geophysical prospection at shallow depths. PhD thesis. University of Bradford, West Yorkshire.

Herbich, T.
2003   Archaeological geophysics in Egypt: The Polish contribution. *Archaeologia Polona* 41:13–55.

REFERENCES

Herbich, T., D. O'Connor, and M. Adams
2003   Magnetic mapping of the northern cemetery at Abydos, Egypt. In "Fifth International Conference on Archaeological Prospection 2003," T. Herbich, ed. *Archaeologica Polona* 41:193–197.

Herbich, T., and C. Peters
2006   Results of magnetic surveying in Deir al-Barsha, Egypt. *Archaeological Prospection* 13 (1):11–24.

Herbich, T., and J. Spencer J.
2006   Geophysical survey at Tell el-Balamun. *Egyptian Archaeology* 29:16–19.

Heron, C. P., and C. F. Gaffney
1987   Archaeogeophysics and the site: Ohm sweet ohm? In *Pragmatic Archaeology: Theory in crisis?* C. F. Gaffney and V. L. Gaffney, eds. Oxford: BAR.

Herwanger, J., H. Maurer, A. G. Green, and J. Leckebusch
2000   Three-D inversion of magnetic gradiometer data in archaeological prospecting: Possibilities and limitations. *Geophysics* 65 (3):849–860.

Hesse, A., L. Barba, K. Link, and A. Ortiz
1997   A magnetic and electrical study of archaeological structures at Loma Alta, Michoacán, Mexico. *Archaeological Prospection* 4 (2):53–68.

Hounslow, M. W., and A. Chepstow-Lusty
2002   Magnetic properties of charcoal-rich deposits associated with a Roman bathhouse, Butrint (Southern Albania). *Physics and chemistry of the earth* 27:1333–1341.

Hrvoic, I., G. M. Hollyer, and M. Wilson
2003   Development of a high-sensitivity potassium magnetometer for near-surface geophysical mapping. *First Break* 21 (May):81–87.

Johnson, R., and K. Smith
2003   Comparing cesium and potassium magnetometers. *First Break*. 21 (Sept.):85–86.

Jones, G.
2001   Geographical investigation at the Falling Creek Ironworks, an early industrial site in Virginia. *Archaeological Prospection* 8 (4):247–256.

Jones, G., and D. L. Maki
2005   Lightning-induced magnetic anomalies on archaeological sites. *Archaeological Prospection* 12 (3):191–197.

Jordanova, N., E. Petrovsky, M. Kovacheva, and D. Jordanova
2001   Factors determining magnetic enhancement of burnt clay from archaeological sites. *Journal of Archaeological Science* 28:1137–1148.

Kastowski, K., K. Loecker, W. Neubauer, and G. Zotti G
2005.   Drehscheibe des Sternenhimmels? Die Kreisgrabenanlage Im-
mendorf. In *Zeitreise Heldenberg: Geheimnisvolle Kreisgraeben—
Niederoesterrichische Landerausstellung 2005.* F. Daim and
W. Neubauer, W., eds. Vienna: KNL,Verlag Berger.

Kattenberg, E., and G. Aalbersberg
2004   Archaeological prospection of the Dutch perimarine landscape by
means of magnetic methods. *Archaeological Prospection* 11 (4):227–235.

Kvamme, K.
2006   Integrating multidimensional geophysical data. *Archaeological Prospection* 13 (1):57–72.

Le Borgne, E.
1955   Susceptibilite magnetique anomale du sol superficiel. *Annales de Geophysique* 11:399–419.
1960   Influence du feu sur proprietes magnetique du sol et du granite. *Annales de Geophysique* 16:159–195.

Li, X., and H. J. Götze
1999   Comparison of some gridding methods. *The Leading Edge* 18 (8):
898–900.

Linford, N. T.
1994   Mineral magnetic profiling of archaeological sediments. *Archaeological Prospection* 1:37–52.
2004   Magnetic ghosts: Mineral magnetic measurements on Roman and
Anglo-Saxon graves. *Archaeological Prospection* 11 (3):167–180.

Linford, N. T., and M. G. Canti
2001   Geophysical evidence of fires in antiquity: Preliminary results from
an experimental study. *Archaeological Prospection* 8 (4):211–225.

Linford, P. K.
2005   An automated approach to the analysis of the arrangement of post
pits at Stanton Drew. *Archaeological Prospection* 12 (3):137–150.

Linington, R. E.
1972–73   A summary of simple theory applicable to magnetic prospecting
in archeology. *Prospezioni Archeoliche* 7-8:9–59.

REFERENCES

Maher, B. A., and R. M. Taylor
1988 Formation of ultrafine-grained magnetite in soils. *Nature* 336: 368–371.

Maki, D., J. A. Homburg, and S. D. Brosowske
2006 Thermally activated mineralogical transformations in archaeological hearths: Inversion from maghemite gamma-Fe_2O_4 phase to hematite alpha-Fe_2O_4 form. *Archaeological Prospection* 13 (3):207–227.

Maki, D., J. A. Homburg, and S. D. Brosowske
2007 Erratum: Thermally activated mineralogical transformations in archaeological hearths: inversion from maghemite gamma-Fe_2O_4 phase to hematite alpha-Fe_2O_4 form. *Archaeological Prospection* 14(1):73.

Marukawa, Y., and H. Kamei
1999 Estimation of the systematic error of three-component magnetic data using the ABIC method. *Archaeological Prospection* 6 (3):135–146.

Mathe, V., and F. Leveque
2003 High-resolution magnetic survey for soil monitoring: Detection of drainage and soil-tillage effects. *Earth and Planetary Science Letters* 212:241–251.

Musty, J. W. G.
1974 Medieval pottery kilns. In *Medieval pottery from excavation: Studies presented to Gerald Clough Dunning, with a bibliography of his works,* V. I. Evisin, H. Hodges, and J. G. Hurst, eds. London: John Baker Publishers.

Neubauer, W.
2001a Images of the invisible-prospection methods for the documentation of threatened archaeological sites. *Naturwissenschaften* 88:13–24.
2001b *Magnetische Prospektion in der Archaologie* Vienna: MPK, Austrian Academy of Sciences.

Neubauer, W., and A. Eder-Hinterleitner
1997 Three-D interpretation of post-processed archaeological magnetic prospection data. *Archaeological Prospection* 4 (4):191–205.

Eder-Hinterleitner, A., P. Melichar, and R. Steiner
2001 Improvements in high resolution magnetometry. In *Filtering, optimization, and modeling of geophysical data in archaeological prospecting,* M. Cucarzu and P. Conti, eds. Milan and Rome: C. M. Lerici Foundation.

Nishimura, Y., M. Saito, and H. Kamei
1996   An applied survey on buried mounded tombs by using three-component magnetometer and earth radar system. *Anales Geophysicae* 14:C164.

Philpot, F. V.
1973   An improved fluxgate gradiometer for archaeological surveys. *Prospezioni Archeologiche* 7-8:99–105.

Powlesland, D., J. Lyall, G. Hopkinson, D. Donoghue, M. Beck, A. Harte, and D. Stott
2006   Beneath the Sand—Remote sensing, archaeology, aggregates, and sustainability: A case study from Heslerton, the Vale of Pickering, North Yorkshire, UK. *Archaeological Prospection* 13(4):291–299

Ralph, E. K.
1964   Comparison of a proton and rubidium magnetometer for archaeological prospecting. *Archaeometry* 7:20–27.

Salvi, A.
1970   Perfectionnements apportes aux magnetometres a resonance magnetique nucleaire a pompage electronique. *Revue de Physique Appliquee* 5:131–134.

Sauerländer, S., J. Kätker, E. Räkers, H. Rüter, and L.Dresen
1999   Using random walk for online magnetic surveys. *European Journal of Environmental and Engineering Geophysics* 3 (2):91–102.

Schleifer, N., J. W. E. Fassbinder, W. E. Irlinger, and H. Stanjek
2003   Investigation of an eneolithic Chamer-group ditch system near Riekofen (Bavaria) with archaeological, geophysical, and pedological methods. In *Soils and Archaeology: Papers of the first international conference on soils and archaeology, Hungary, 2001,* G. Füleky, ed.Oxford: Archaeopress.

Schmidt, A.
2001   Visualization of multi-source archaeological geophysics data. In *Filtering, optimization, and modeling of geophysical data in archaeological prospecting,* M. Cucarziand, P. Conti, eds. Milan and Rome: C. M. Lerici Foundation.
2002   *Geophysical data in Archaeology: A guide to good field practice.* Oxford: Archaeological Data Service and Oxbow Books Books. http://ads.ahds.ac.uk/project/goodguides/geophys/

REFERENCES

2003   Remote Sensing and Geophysical Prospection. In *Internet Archaeology 15*. http://intarch.ac.uk/journal/issue_15/schmidt_index.html.

Schmidt, A., and A. Marshall
1997   Impact of resolution on the interpretation of archaeological prospection data. In *Archaeological Sciences 1995*, A. Sinclair, E. Slater, and J. Gowlett, eds. Oxford: Oxbow Books.

Schmidt, A., and H. Fazeli
2006   Tepe Ghabristan: A Chalcolithic tell buried in alluvium. *Archaeological Prospection* 14 (1):38–46.

Schmidt, P. W., and D. A. Clark
2006   The magnetic gradient tensor: Its properties and uses in source characterization. *The Leading Edge*. 25 (1):75–78.

Schultze, V., A. Chwala, R. Stole, M. Schulz, S. Linzen, H.-G. Meyer, and T. Schüler
2007   A SQUID system for geomagnetic gradiometry. *Archaeological Prospection*.

Scollar, I.
1969   A program for the simulation of magnetic anomalies of archaeological origin in a computer. *Prospezioni Archeologiche* 4:59–83.

Scollar, I., D. Gubbins, and P. Wisskirchen
1971   Two-dimensional digital filtering with Haar and Walsh transforms. *Annales Geophysicae* 27: 85–104.

Scollar, I., A. Tabbagh, A. Hesse, and I. Herzog
1990   *Archaeological prospecting and remote sensing*. Cambridge: Cambridge University Press.

Sheen, N.
1997   Automatic interpretation of magnetic gradiometer data using a hybrid neural network. PhD thesis Bradford University, West Yorkshire.

Shell, C. A.
1996   Magnetometric surveys at Catalhoyuk East. In *On the surface: Catalhoyuk 1993–95*, I. Hodder, ed. Cambridge: McDonald Institute for Archaeological Research and British School of Archaeology at Ankara.

Smith, N., F. Jeffers, and J. Freeman
1991   A high-sensitivity magnetoresistive magnetometer. *Journal of Applied Physics* 69 IIA:5082–5084.

Somers, L., M. L. Hargraves, and J. E. Simms
2003   *Geophysical surveys in archaeology: Guidance for surveyors and sponsors.* U.S. Army Corps of Engineers, Engineering Research and Development Center. ERDC/CERL SR-03-21. www.cecer.army.mil/.

Sowerbutts, W. T. C.
1988   The use of geophysical methods to locate joints in underground metal pipelines. *Quarterly Journal of Engineering Geology* 21:273–281.

Sutherland, T. L.
2007   Arrows point to mass graves: Finding the dead from the battle of Towton, 1461. In *Fields of conflict: Battlefield archaeology from the Roman Empire to the Korean War* (1), D. D. Scott, L. Babits, and C. Haeker, eds. Westport, Connecticut: Praeger Security International.

Sutherland, T. L., and A. Schmidt
2003   Towton 1461: An integrated approach to battlefield archaeology. *Landscapes* 4 (2):15–25.

Tabbagh, J.
2003   Total-field magnetic prospection: Are vertical gradiometer measurements preferable to single-sensor survey? *Archaeological Prospection 10* (2):75–81.

Tabbagh, A., G. Desvignes, and M. Dabas
1997   Processing of Z gradiometer magnetic data using linear transforms and analytical signal. *Archaeological Prospection* 4 (1):1–14.

Tarling, D. H.
1983   *Palaeomagnetism: Principles and applications in geology, geophysics, and archaeology.* London and New York: Chapman and Hall Publishing.

Telford, W. M., G. P. Geldart, and R. E. Sheriff
1990   *Applied Geophysics.* Cambridge University Press: Cambridge.

Tite, M. S,
1966   Magnetic prospecting near the geomagnetic equator. *Archaeometry* 9:24–31.
1972   The influence of geology on the magnetic susceptibility of soils on archaeological sites. *Archaeometry* 14: 229–236.

REFERENCES

Tite, M. S., and C. Mullins
1970   Magnetic properties of soils. *Prospezione Archeologiche* 5:111–112.
1971   Enhancement of the magnetic susceptibility of soil on archaeological sites. *Archaeometry* 13 (2): 209–220.

Tite, M. S., and R. E. Linington
1975   Effect of climate on the magnetic susceptibility of soils. *Nature* 256:565–566.
1986   The magnetic susceptibility of soils from central and southern Italy. *Prospezione Archeologiche* 10:25–36.

Tsokas, G. N., and R. O. Hansen
1995   A comparison of inverse filtering and multiple-source Werner deconvolution for model archaeological problems. *Archaeometry* 37:185–193.

Vehik, S. C.
2002   Conflict, trade, and political development on the Southern Plains. *American Antiquity* 67 (1): 37–64.

Vernon, R. W., J. G. McDonnell, and A. Schmidt
1998   The geophysical evaluation of an iron-working complex: Revaulx and environs. *Archaeological Prospection* 5 (4):181–201.
2002   The geophysical evaluation of British lead- and copper-working sites: Comparison with ironworking. *Archaeological Prospection* 9 (3): 123–134.

Walker, A. R., C. F. Gaffney, J. A. Gater, and E. Wood
2005   Fluxgate gradiometry and square-array resistance survey at Drumlanrig, Dumfries, and Galloway, Scotland. *Archaeological Prospection* 12 (2):131–136.

Waters, G. S., and P. D. Francis
1958   A nuclear magnetometer. *Journal of Scientific Instruments* 35:88–93.

Weston, D. G.
1995   A magnetic susceptibility and phosphate analysis of a longhouse feature on Caer Cadwgan, near Cellan, Lampeter, Wales. *Archaeological Prospection* 2 (1):19–27.
1996   Soil science and the interpretation of archaeological sites: A soil survey and magnetic-susceptibility analysis of Altofts henge, Normanton, West Yorkshire. *Archaeological Prospection* 3 (1):39–50.
2001   Alluvium and geophysical prospection. *Archaeological Prospection* 8 (4):265–272.

2002   Soil and susceptibility: Aspects of thermally induced magnetism within the dynamic pedological system. *Archaeological Prospection* 9 (4):207–215.

2004   The influence of waterlogging and variations in pedology and ignition upon resultant susceptibilities: A series of laboratory reconstructions. *Archaeological Prospection* 11 (2):107–120.

Wheeler, J., A. Aspinall, and A. R. Walker

2007   Geophysics in the garden: A survey of the garden at Duncombe Park, North Yorkshire. *Garden History* 35 (1):85–91.

White, R., and P. Barker

1998   *Wroxeter: Life and death of a Roman city*. Stroud, Gloucestershire: Tempus Publishing.

Weymouth, J. W., and Y. A. Lessard

1986   Simulation studies of diurnal corrections for magnetic prospection. *Prospezioni Archeologiche* 10:37–47.

INDEX

ABOUT THE AUTHORS

Arnold Aspinall developed an interest in the application of geophysical techniques for archaeological survey in the early 1960's. Over the succeeding years, he supervised a number of doctorate projects at the University of Bradford relating to the development of magnetic, electromagnetic and earth-resistance procedures for rapid and effective survey. He initiated the first postgraduate and undergraduate teaching programs in archaeological science in the United Kingdom with specific modules and, later, complete courses dealing with "archaeological prospection," and became a foundation editor of the international journal with that same title. Since his retirement from academic life in 1991, he has maintained an active interest in advancing techniques now employed in archaeological survey. He is an Honorary Professor in Archaeological Sciences at Bradford and has an honorary doctorate from Sheffield University. He was also among the first group of honorary members of the International Society for Archaeological Prospection (ISAP).

Chris Gaffney worked for eighteen years in commercial archaeology as the co-owner of GSB Prospection. He has a Ph.D in earth-resistance and has published in many areas of archaeological prospecting. As editor of *Archaeological Prospection* and holder of committee positions within ISAP, he has many links with the geophysical community, while his work popularizing archaeological geophysics was acknowledged by the awarding of an honorary D.Sc in 2006. Recently he has returned to the academic world, and is a lecturer at the University of Bradford. His interests include rapid collection of data rich data sets, particularly at the landscape level.

Armin Schmidt has a Ph.D in physics but was so captivated by Helmut Becker's guest lecture on the use of magnetometer surveys for archaeology, that he decided to become an archaeological geophysicist. He is now leading the Archaeological Prospection Research Group at the University of Bradford where he is Senior Lecturer. As manager of the M.Sc in Archaeological Prospection he has supervised over thirty M.Sc dissertations and several doctorates. He is interested in all fields of archaeological prospection and combines instrument and method development with archaeological research (ranging from British prehistoric sites, Iranian chalcolithic tells, Nepalese temples to Australian gardens).